ESSENTIALS OF BACTERIAL AND VIRAL GENETICS

Essentials of

Bacterial and

Viral Genetics

By

DAVID M. CARLBERG, Ph.D.
Professor
Department of Microbiology
California State University
Long Beach, California

CHARLES C THOMAS · PUBLISHER
Springfield · Illinois · U.S.A.

Published and Distributed Throughout the World by
CHARLES C THOMAS · PUBLISHER
Bannerstone House
301-327 East Lawrence Avenue, Springfield, Illinois, U.S.A.

*With THOMAS BOOKS careful attention is given to all details of
manufacturing and design. It is the Publisher's desire to present books that are
satisfactory as to their physical qualities and artistic possibilities and
appropriate for their particular use. THOMAS BOOKS will be true to those
laws of quality that assure a good name and good will.*

Library of Congress Cataloging in Publication Data

Carlberg, David M
  Essentials of bacterial and viral genetics.
  Bibliography: p. 299.
Includes index.
  1. Microbial genetics. I. Title, [DNLM: 1. Genetics, Microbial. 2. Viruses. QW51 C278e]
QH434.C37   589.9'01'5   75-23269
ISBN 0-398-03493-1

Printed in the United States of America
M-3

This volume is warmly dedicated to my wife, Margaret, and the boys, Howard and Marvin, who silently endured its long gestation period.

. . . And to the memory of Anna May Brodetsky

PREFACE

Microbial genetics continues to play an increasingly larger role in modern biological research. In every issue of the world's scientific journals one finds significant numbers of papers involving applications of microbial genetics. In addition to aiding us in the understanding of basic biological mechanisms, aspects of microbial genetics also help us to explain such features of immediate concern as phage typing patterns and antibiotic resistance among microorganisms of clinical interest, and pathways of industrial fermentations. As an even more narrow example, the fundamentals of genetic engineering now being developed from work on microorganisms will some day be practiced on higher forms, including man, leading to the eventual treatment of scores of inborn diseases. Thus, the study of the genetics of microorganisms is as important to students entering careers in medical and industrial microbiology and medicine, as it is to the researcher and the teacher.

This volume is intended to be used as a textbook for an undergraduate course in microbial or bacterial genetics. It was prepared with the assumption that its users would have had at least one course each in microbiology and organic chemistry. The book strikes at the essentials of bacterial and viral genetics, leaving much fine detail for the interested student to seek out in the literature. One of the most difficult decisions in this connection has been the selection of the references that appear at the end of the text. The mass of references in microbial genetics and molecular biology obviously is too enormous to include in an undergraduate textbook. Consequently, a careful selection of references appears on the pages that follow. Many of the citations are for review articles such as those that appear in *Annual Review of Microbiology* and *Current Topics in Microbiology and Immunology*. Many references cite original work, however, to enable students to find readily those papers of

vii

particular historic value or those that describe a useful technique or support an important hypothesis.

The indispensable desk reference for years has been *The Genetics of Bacteria and Their Viruses* by William Hayes (2nd edition, 1968, Wiley, New York). Hayes has also coauthored the excellent laboratory manual *Experiments in Microbial Genetics* with R. C. Clowes (1968, Wiley, New York). More advanced techniques are described in *Experiments in Molecular Genetics* by J. H. Miller (1972, Cold Spring Harbor Laboratory, New York). A strain kit of bacteria and viruses that is coordinated with the manual is also available from this publisher. A concise collection of experiments will be found in *Genetics Experiments with Bacterial Viruses* by D. P. Snustad and D. S. Dean (1971, Freeman and Company, San Francisco). Readers will find *Molecular Genetics, an Introductory Narrative,* by Gunther Stent (1971, Freeman and Company, San Francisco) a valuable adjunct to any discussion of microbial genetics.

Every few years collections of important original papers appear in volume form. One significant instance is *Classic Papers in Genetics*-by J. A. Peters (1959, Prentice-Hall, Inglewood Cliffs, N. J.). Its importance lies in the fact that it contains an English translation of Mendel's original paper, as well as over two dozen other milestone papers. Other collections are: *Papers on Bacterial Genetics,* by E. A. Adelberg, and *Papers on Bacterial Viruses,* by G. Stent, both published in 1960 by Little, Brown and Company, Boston. A valuable collection of papers will be found in *Selected Papers on Molecular Genetics,* by J. H. Taylor (1965, Academic Press, New York). A more recent volume of reprints is edited by M. A. Abou-Sabe: *Benchmark Papers in Microbiology: Microbial Genetics* (1973, Dowden, Hutchinson and Ross, Stroudburg, Pa.).

A major part of this book was written during a sabbatical leave from California State University, Long Beach, for which I am deeply grateful.

Sincere acknowledgements are made to Doctors Anna May Brodetsky and Herbert Eichhorn for their invaluable criticisms and suggestions. Many thanks to Cheryl Nau for her typing and editorial skills, and to Kim Odenweller and Thom Jacobs for lending their talents for many of the illustrations. Special appreciation must go to my wife for reading and rereading the manuscript, and for typing some of the more arduous portions of the final draft. The sole responsibility for errors that may have slipped past these very able people is, of course, mine.

I am also grateful to the many authors and publishers who gave their permissions for using their illustrations and data. They are cited individually in appropriate locations.

David M. Carlberg

CONTENTS

ESSENTIALS OF BACTERIAL AND VIRAL GENETICS

Chapter One

INTRODUCTION

THE FOUNDATIONS OF
BIOLOGY AND GENETICS

THE EXISTENCE OF creatures too small to be seen with the unaided eye had been hypothesized at least as far back as the sixteenth century, but their true appearance was not made known to us until the invention of the microscope sometime in the seventeenth century. Prominent naturalists such as Athanasium Kircher (1602-1680) and Robert Hooke (1635-1703) are generally recognized as being the first to observe microorganisms through the microscope. The man usually given credit for being the first to describe microorganisms in great detail was the Dutch shopkeeper and amateur naturalist Antonie van Leeuwenhoek. From about 1660 until well into the 1700's, Leeuwenhoek spent his spare time examining rain water, urine, and other specimens with his handcrafted microscopes and describing in numerous letters and papers what he saw. In spite of the simplicity and crudeness of his instruments, his drawings of protozoa and bacteria were incredibly accurate, and they created a sensation among his scientist contemporaries. He is frequently referred to as the father of bacteriology and protozoology.

The role of microorganisms in the preparation of certain foods, and in their spoilage as well, was established by the French chemist-turned-microbiologist, Louis Pasteur, during the years following about 1855. Pasteur's studies of the causes of the spoilage of liquids such as beer, wine, and beef broth led to the raging controversy regarding spontaneous generation.

With the correct ingredients (old rags for mice, meat for maggots, or beef broth for microbes) set aside for some days, life would spontaneously spring from them, according to the theory.

The genesis of the theory dates back to ancient times, and Pasteur barely a century ago ended it with a series of simple experiments. He showed that if a sterile, fermentable liquid is protected from particulate fall-out, the liquid remained unspoiled. Leave the liquid open to dust, and it soon becomes turbid with microbial life.

That microorganisms were responsible for the incidence of contagious disease was hypothesized by the sixteenth century scientist Fracastoro and exhaustively proved by the German Robert Koch in 1876. That there were yet smaller creatures responsible for infectious diseases, that is, viruses, was first shown by Iwanowski, a Russian botanist, in 1892. Thus, by the turn of the twentieth century, microbiology was a firmly established branch of biology.

Widespread use of microscopes in the early nineteenth century led to the establishment of one of the most significant principles in biology, the cell theory of Schleiden and Schwann. In it they proposed in 1839 that all plants and animals are composed of discrete units called *cells*. It should be noted that the pioneer microscopist Robert Hooke recognized the existence of cells in plants in 1665 and in fact coined the term *cell*. Cells occur in a vast array of sizes, from 0.001 mm (1 µm) for a typical bacterial cell, to 80 mm (80,000 µm) for an ostrich ovum. However, they all have features in common: They all experience an exchange of matter and energy to sustain their life, and they all have a means of transmitting information to subsequent generations of cells in order to perpetuate their kind.

There is no record of man's first recognition that like creatures spring from like. Mice produce only mice, camels only camels. But the likenesses were never absolute; progeny were seldom exactly like the parents, sibs seldom identical. This discovery was also prehistoric, for biblical writings and other sources indicate that plant and animal hybridization (the bringing together of nonidentical but desirable traits through breeding) was well established by perhaps 6000 B.C.

Thus, ancient scholars (and farmers) *knew* that living things transmitted their likenesses to their offspring, but the mechanism, the *how,* would not be known to us until the nineteenth century. Improvements in microscopes made it possible to discern the inner details of cells, and in 1831 the Scottish biologist Robert Brown discovered a prominent structure that appeared to be in all cells which he named the *nucleus.* In 1855 the German botanist Nathanael Pringsheim observed that on conjugation of the gametes of the alga *Vaucheria,* the nuclei appeared to fuse and become one in the *zygote.* An identical observation was made by Hertwig in the fertilization of sea urchin eggs in 1865. The nucleus must play a role in the transmission of hereditary information was the conclusion of many biologists by the 1880's, but the mechanism of transmission still eluded them. The answer was at hand, discovered by an Austrian monk in the 1860's, but was not to be recognized until the turn of the century.

During these later years of the nineteenth century there appeared in a narrowly distributed journal a paper by an Austrian monk, Gregor Mendel, describing eight years of experiments involving the formation of hybrids of garden peas. Mendel had always been interested in plants and plant breeding. He grew up in a farming community, and his father had a small fruit orchard. After having entered the Augustinian order, Mendel took some university courses in science in order to qualify as a teacher. He did poorly, however, and was forced to return to the monastery and to a life of much spare time during which he could pursue his lifetime interest, plant breeding.

Mendel's paper, published in 1865 in the Proceedings of the Brünn Natural History Society, reported that certain predictable mathematical patterns appeared whenever hybrids were formed of plants of certain differing characteristics. The author outlined some general laws that govern the formation of hybrids that later proved to be significant milestones in the history of science. But the paper remained largely unnoticed and unappreciated for thirty-five years, until in 1900 three botanists working independently in three countries came across Mendel's report. The three, Carl Correns of Germany, Hugo de Vries of Holland, and Erich von Tschermak of Austria immediately recognized the great significance of Mendel's work. The practice of hybridization had been known for at least 6000 years, but Mendel was the first to follow analytically the inheritance patterns of simple, easily recognizable traits and thereby establish the foundations of modern genetics.

These foundations, known as Mendel's Laws of Inheritance, include the Law of Segregation and the Law of Independent Assortment.

Law of Segregation

This principle states that inherited traits in plants and animals are controlled by pairs of factors or elements (now loosely called *genes*) in the cells. On formation of *gametes* (sperm or ova), the pairs segregate so that each sperm or ovum contains but one member of each pair. On fertilization of an ovum by a sperm cell, the resulting *zygote* now has the pairs of factors restored, one member received from the sperm, one that was in the ovum.

Figure 1 illustrates a typical cross that Mendel carried out in which one strain of pea having tall stems was crossed with one with short stems. Mendel visualized that the tall plants had the heredity element pair TT in their cells, the short had the pair tt. Thus, each pollen of the tall plant would contain one T, each ovum of the short, one t. On fertilization each resulting zygote would then consist of the pair Tt, and as Mendel observed, every resulting plant would be tall in spite of the presence of the "short" element in the plants. Mendel concluded that the tall element T was *dominant* over the short element t, which he referred to as being *recessive*. If this (F_1) generation was then crossed with itself, some of the resulting F_2 progeny were tall and some were short, but it was the numbers of short plants compared to the tall that attracted Mendel's attention. Re-

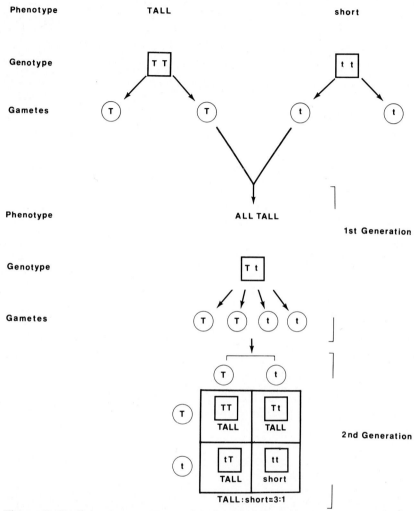

Figure 1. Outline of one of Mendel's crosses with tall and short pea plants. Genotypes with respect to height of mature plants are shown in boxes, while those of gametes (pollen or ova) are in circles.

peatedly Mendel observed the ratio approximately 3:1 to the alternate recessive trait. Thus, Mendel explained, the tall plants had either a TT pair or a Tt; the small had to have a tt pair.

Law of Independent Assortment

The Law of Independent Assortment deals with the behavior of the trait-controlling elements when studied in groups of two or more. That is, what happened when Mendel crossed tall plants having smooth, grey seeds with short plants having wrinkled, white seeds? The same basic ratios prevailed for each trait, independent of the other traits. Mendel concluded the elements segregated and distributed themselves among the generations in a manner completely independent of one another. The Austrian monk was not concerned with the physical nature of his factors, only wih their movements from generation to generation. This does not detract from the remarkable accomplishment of this amateur scientist, an accomplishment that is marred only by the fact that its true significance was not recognized until sixteen years after Mendel's death.

The turn of the century saw groups of biologists forming at institutions such as Columbia University and California Institute of Technology in the United States and Cambridge in England to begin studies of the nature of inheritance in plants and animals. one of the prominent members of the Cambridge group was William Bateson, who coined the term "genetics" in 1905 and did much to stimulate interest in this new science.

Chromosome Theory of Inheritance

Much of what was learned during the early 1900's supported the basic laws set down by Mendel, but many contradicted them. For example, certain combinations of traits did not follow the law of independent assortment. The possibility that this could occur had been predicted some four years before it was observed by a Columbia graduate student, Walter Sutton. Sutton, as any good graduate student, was fully aware of the literature. He knew of the work of Hertwig and Strasburger and of course, Mendel. Of greater importance to Sutton was the behavior of the deeply staining, threadlike bodies in the nucleus, the *chromosomes*. Every cell of a plant or animal had a complement of chromosomes, the exact number being the same in the same species but differing widely among different species. On division of a cell the chromosomes appeared to divide longitudinally, with one half going to each daughter cell in a remarkable ballet known as *mitosis*.

Thus, if the chromosomes carried genetic information, a suggestion made as early as the 1880's, then each daughter cell would receive a complete set of information. Were the chromosomes Mendel's "elements?" Probably not, for the size and complexity of an animal has no apparent relation to its number of chromosomes. Houseflies have 12, man 46, but goldfish have 94 and certain crustaceans, over 200. Sutton concluded then that the chromosomes were made up of individual genes of the organism, and it would be very likely that if two traits one were studying happened to be on the same chromosomes, the traits would not demonstrate Mendel's independent assortment.

PRINCIPLES OF TWENTIETH CENTURY GENETICS

The science of genetics developed to a high degree during the first half of the twentieth century. Examples of some of the significant principles that were established during this period are:

a. GENES ARE CARRIERS OF GENETIC INFORMATION—this concept was formulated before 1900 and strengthened by Sutton's chromosomal theory of inheritance. The *genotype* of an organism is the sum total of an organism's genetic information, and it is the genotype that controls the organism's outward appearance, or *phenotype*. This is not to imply that the genotype is the only controlling force over the phenotype, for the environment can also influence the latter. The difference is, the effects of the environment are not permanent and are not transmitted to subsequent generations, while those of the genotype are. The establishment of this notion probably was responsible for more bitter debate than any other concept we will deal with, particularly as it relates to microorganisms.

b. GENES ACT ON BIOSYNTHETIC PATHWAYS—the great chemist Berzelius reflected in 1836 that thousands of catalytic processes take place in living plants and animals to form like numbers of compounds, and he proposed that such catalytic power would prove to be the very organic tissue of which the plants and animals are composed. Berzelius thus predicted the existence of enzymes, the protein catalysts that drive the metabolism of the cell. By the turn of the century, biochemistry had established significant foundations in the study of the nature and functions of the chemicals associated with living cells. If biochemical reactions were the ultimate phenotype, what was the exact nature of the control of these reactions by the genotype?

A. E. Garrod, an English physician, is usually credited with casting the first light on the relationship between the genotype and metabolic pathways. Garrod observed that persons suffering from certain diseases such as alkaptonuria were frequently the children of first cousin marriages. The incidence of the disease followed a pattern one would expect if it were controlled by a simple Mendellian recessive gene. The defect itself, Garrod discovered, was due to the inability of the person to convert the aromatic amino acid phenylalanine into tyrosine. The enzyme responsible for this reaction is lacking, no doubt due to a faulty gene. Garrod studied other metabolic diseases and arrived at similar conclusions. His work was summarized in a book, *Inborn Errors of Metabolism,* published in 1903, and also in a paper in the British medical journal *Lancet* in 1908, but both were virtually ignored for over thirty years.

Through the years attempts to establish experimentally a connection between genes and biochemical pathways generally were fruitless, until a geneticist at California Institute of Technology, George Beadle, began to accumulate good experimental evidence that the action of genes was the control of the formation of the enzymes that catalyzed metabolic reactions. The system that Beadle first studied was the formation of eye pigments in the fruit fly, *Drosophila. Drosophila* had become a favorite object of genetic research. Though not ideal, its generation time and fecundity were such that an experiment could be run in perhaps eight to ten weeks. But its physiology was not well known, and Beadle became frustrated in his inability to learn the detailed pathways leading to the pigments and turned instead to a less complex organism. *Neurospora,* an ascomycete, had been the object of some genetic studies in the early 1930's. It grew on a chemically defined medium, and of course exhibited the prime microbial trait of a very short generation time. Beadle had moved to Stanford and was joined by Edward Tatum, a biochemist, to begin work which was to prove to be some of the most significant in genetics. Beadle and Tatum managed to isolate strains of the original fungus that required additional growth factors to be added to the medium such as paraaminobenzoic acid or pyridoxine. The organism, normally capable of synthesizing these factors, had presumably lost this ability. Beadle and Tatum were able to associate the loss with the absence of a single enzyme, which in turn was under the influence of a single gene. From this work they formulated the *one gene-one enzyme hypothesis,* which states that for every enzyme in a cell there is a particular gene which is responsible for the formation of that enzyme. While this concept paraphrases what Garrod had tried to show, this was the first time substantial experimental evidence had been presented to support it. It is also the first time that a microorganism was put to extensive use in the study of genetics.

    c. GENETIC INTERACTIONS—one of Mendel's most significant observations was that his pea plants did not exhibit a blending of traits. The progeny resulting from a cross between long-stemmed plants and short-stemmed plants did not have stems of intermediate length. However, it was soon demonstrated that some genes do interact to result in intermediate forms. Such an example is *multiple factor genes,* or *polygenes,* in wheat, where there are two pairs of genes responsible for seed color. Depending on the distribution of the recessive and dominant genes, a plant may exhibit seeds with any of four shades of red, or white (Fig. 2). Penicillin resistance in bacteria is a similar example in which several genes are associated with resistance, and the phenotype exhibited is the result of the additive effects of the genes.

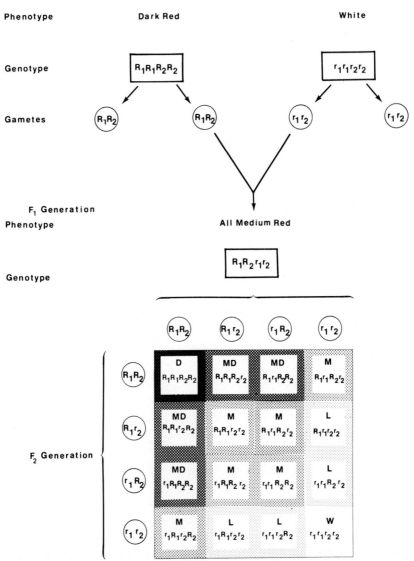

Figure 2. Polygenes in wheat. When plants with dark red seeds are crossed with plants with white seeds, all of the resulting progeny plants produce medium red seeds. When these are sown and the resulting plants are crossed with themselves, the next generation exhibit seeds of four shades of red and white. Two distinct genes are responsible for seed color, designated as R_1 and R_2 for the dominant forms of the alleles, and r_1 and r_2 for the recessive forms. (D = dark red, MD = medium dark, M = medium red [like the F_1 generation], L = light red, and W = white. (From *Principles of Genetics* (5th edition) by Sinnott, E. W., Dunn, L. C., and Dobzhansky, T. New York, McGraw, 1958)

Another interaction of genes of a different sort is seen in the case of pleiotropic genes. A *pleiotropic gene* is one that controls two or more seemingly unrelated phenotypic traits. In the fruit fly, *Drosophila,* one might find a single gene that appears to control eye color, wing shape, wing vein pattern, body hair, and fertility. This is not a contradiction of the one gene-one enzyme hypothesis, for it appears that the traits mentioned probably all share a common factor under control of the single pleiotropic gene.

d. MUTATIONS—undoubtedly one of the most salient discoveries in the history of genetics was the finding soon after the turn of the century of the existence of mutations. Mutations are randomly occurring changes in the genotype of an organism. We now know they are the result of a change in the structure of the molecule carrying genetic information, and they are capable of being transmitted to the progeny of the cell that experienced them. We will devote a considerable amount of space discussing the nature of mutations in Chapters Four and Five. Many mutations probably go undetected, whereas others are observable as losses of certain functions, such as the formation of vitamins in the case of the *Neurospora* studies of Beadle and Tatum. Mutations occur spontaneously in all organisms, the rate of occurrence being apparently constant and measurable for a given gene. For example, mutations in *Staphylococcus* that afford the bacterium resistance to penicillin occur at a rate of about one mutation per generation per 10^6 cells. Loss of pigment formation in *Serratia* as a result of a mutation occurs about once per generation per 10^4 cells.

In 1927 the American biologist Muller showed that if he exposed *Drosophila* fly larvae to X radiation, he could increase the rate at which mutations occurred by factors of 100 or more over the spontaneous rates. This was the first instance when man was able to intervene in a genetic process. Other sources of ionizing radiation, such as that from radioactive elements, are also capable of inducing mutations. In the 1940's, certain chemicals were shown to be mutagenic, as well.

It was realized early in the twentieth century that mutations must be one of the mechanisms by which evolution operates. As environments change over the eons, those organisms that had assembled a genotype through mutations (and through recombination, which is discussed in Chapter Six) that would afford them greater success in the new environment would dominate, but these might be replaced at some later time by yet fitter organisms.

The study of the genetics of animals and plants had expanded to a high level of activity by the 1940's. A wide array of experimental organisms were employed during this period: primroses, corn, peas, rabbits, mice, guinea pigs, and of course the tiny fruit fly, *Drosophila*. Experiments generally required the use of hundreds, thousands, and at times, tens of thousands of individuals. It soon became clear that there were no laboratories large enough to house the numbers necessary to carry out some of the experiments planned. Even *Drosophila*, as small as it was, demanded large amounts of space and manpower. What was needed was an experimental organism that had a chemically defined diet, occupied very little space, demonstrated easily counted phenotypic traits, and enjoyed a short generation time. Microorganisms would appear to fill these requirements, and of course Beadle and Tatum and others had already begun genetic studies with the fungus *Neurospora*. *Neurospora* had its shortcomings, however. To determine the results of a cross between two mutants of *Neurospora*, one usually had to separate the ascospores from the ascus resulting from the cross with a microneedle and deliver each to an appropriate medium to test its phenotype. But what of bacteria? They would appear to fulfill these requirements quite satisfactorily—growth on glucose and mineral salts medium, thirty-minute generation time, and one could comfortably accommodate 10^9 individuals in a single test tube!

The problem with bacteria, as was pointed out prior to 1940, was that there was no evidence for the presence of genes or chromosomes or any genetic apparatuses in them. They were thought to be bags of enzymes which, being so closely associated with the environment, were easily modified by it. This concept of the active role of the environment in the evolution of organisms was proposed by the French biologist Lamarck in the late eighteenth century. Variations, he suggested, which we observe in animals, are the result of changes which are induced in response to some pressure brought about by the environment. These changes, Lamarck concluded, were heritable. The Lamarckian doctrine as it applied to higher plants and animals was readily dismissed by the turn of the century, but in spite of the fact that bacteria were intensely studied for over fifty years, the Lamarckian philosophy remained with them until the 1950's. Once this doctrine was found no longer to apply to bacteria and microorganisms in general, microbial genetics was firmly established and, in fact has proved to be an unmatched source of understanding of the mechanics of heredity.

HISTORY OF GENETIC MATERIAL

DISCOVERY

NO SOONER HAD IT BEEN observed that all living organisms consist of cells than the chemical nature of the cells was investigated. During this period of time (early 1800's) there had been a widespread notion known as *vitalism* which stated that all organic compounds could be synthesized only by living cells. The cells possessed a vital force that could not be duplicated in a test tube. Then, in 1828 the German chemist, Friedrich Wöhler, managed quite by accident to synthesize urea from inorganic salts, and the vitalism concept immediately began to lose support. By the 1850's a number of organic compounds had been synthesized from inorganic starting material. Synthetic organic chemistry led to the development of quantitative elemental analysis, which ultimately enabled chemists to discover the composition of the more complex constituents of living cells, lipids, proteins, carbohydrates and nucleic acids.

Of the major macromolecules in cells, the nucleic acids are of greatest interest to geneticists. A Swiss chemist, Friedrich Miescher, turned to research in histochemistry after graduating from medical school in 1868 and joined Hoppe-Seyler, one of the founders of biochemistry. Miescher chose human leucocytes as the first object of his investigations; he had a nearly unlimited supply of them in the discarded bandages from the University of Tübingen surgical clinic. The nuclei of the pus cells attracted his attention, and he found he could isolate them by subjecting whole cells to proteolytic enzymes and hydrochloric acid. The cell walls and cytoplasm dissolved away, leaving the nuclear bodies. Additional precipitations and dissolutions with acid and alkali, respectively, led to the isolation of a gelatinous material from the nuclei Miescher named *nuclein*. Chemical analysis revealed nothing unusual about the material except its high phosphorus content.

Miescher moved to the University of Basel, which happens to overlook the headwaters of the Rhine River. The salmon which filled the waters of the river were heavily laden with sperm and eggs, and the sperm, as Miescher was to learn, are over 90 percent nucleus, an excellent source of nuclein. Miescher continued his work on nuclein until his death in 1895, and others carried it further. One of them, Richard Altmann, renamed the phosphorus-containing component of nuclein *nucleic acid* in 1889.

The function of nucleic acid eluded most biochemists of the time. Miescher felt it played a significant role in fertilization, but what, he would not speculate. The biologist, Hertwig, had spent decades studying fertilization and was willing in 1884 to propose that the nucleic acid was responsible for the transmission of hereditary information. Others made the same proposal, and by 1900 most biologists agreed that nucleic acid was synonymous with Mendel's "elements."

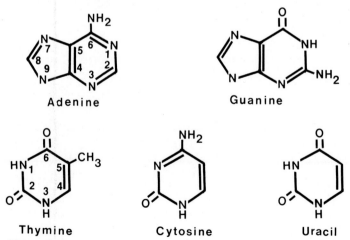

Figure 3. Structures of the common purine and pyrimidine bases.

Most of the major constituents of nucleic acid were known by 1900 and consisted of five purines and pyrimidines (Fig. 3). A five-carbon sugar, ribose, was found in 1909, and not until 1930 was a second sugar discovered in nucleic acid, deoxyribose (Fig. 4). By now it was clear that there were actually two kinds of nucleic acid, one that contained ribose and is now referred to as ribonucleic acid, or RNA, and one that contained the deoxyribose, referred to as deoxyribonucleic acid, or DNA.

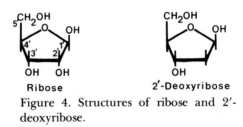

Figure 4. Structures of ribose and 2'-deoxyribose.

Ribonucleic acids generally contain the pyrimidines cytosine and uracil, deoxyribonucleic acids cytosine and thymine. Both contain adenine and guanine. In later chapters we will have the opportunity of learning of the existence of other minor bases of nucleic acids, such as 5-methylcytosine and 1-methylguanine. The third component of nucleic acids, phosphate, was of course originally discovered in them by Miescher.

At first it appeared that RNA occurred only in yeast, while DNA was found only in animals, particularly the thymus gland. Hence, in the older literature one frequently sees reference to yeast nucleic acid and thymus nucleic acid. Not until the 1940's was it established that in actuality both kinds of nucleic acids are found in all cells. DNA is found predominantly in the nuclei of cells and was the type of nucleic acid discovered by Miescher. Mitochondria, kinetoplasts and chloroplasts also contain DNA. RNA, on the other hand, is found almost exclusively in the cytoplasm.

BIOLOGICAL ROLE AND CHEMICAL NATURE OF DNA

During the early decades of this century, the key hereditary role assigned to DNA by biologists before the turn of the century began to lose support from two forces. First, the amount of DNA in cells as determined by staining appeared to vary according to the stage of cell division. Genetic information could not wax and wane in this manner. In addition, protein, a second substance found in relatively large quantities in nuclei, had by now been analyzed and found to consist of over twenty different building blocks, the amino acids. DNA, on the other hand, consists of only four. Obviously, such a simple molecule as DNA could not be capable of carrying the enormous and complex fund of genetic information of an organism. Protein no doubt could, and by the 1920's it was concluded that DNA was the simple molecular scaffold on which proteinaceous genes hung.

The biological role of DNA was thus relegated to a passive one in the early years of this century, and biochemists turned their attention to the apparent central molecules of living cells the proteins.

Two events occurred toward the end of the 1920's which marked the beginning of the reestablishment of the central position of DNA. The first was an observation by Gates (1928) that the action spectrum of ultraviolet (UV) light's effect on bacteria (wavelength vs. number of cells killed) coincided more closely with the absorption spectrum of the constituents of nucleic acids than with the spectra of the UV absorbing constituents of protein (Fig. 5).

The second observation, also in 1928, was made by the British bacteriologist Griffith. He was concerned with capsular types among the pneumococci and found he could change the capsular type of certain strains by some laboratory manipulations. The virulence of the pneumococcus is enhanced by the formation of a polysaccharide capsule. The presence of the capsule imparts a smooth appearance to the colonies of the virulent strain. Absence of the polysaccharide affords a dull, rough colonial morphology and relative avirulence towards

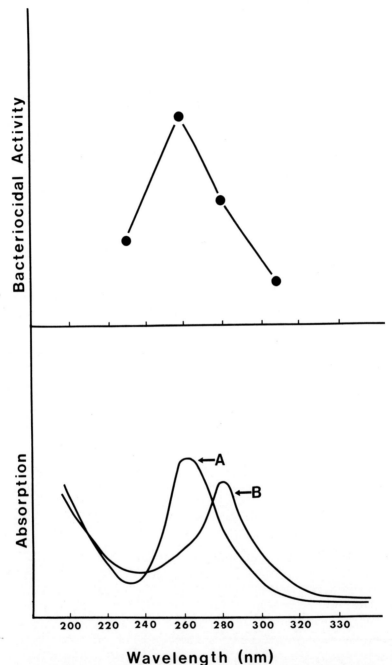

Wavelength (nm)

Figure 5. Bacteriocidal action spectrum of ultraviolet light. Gates (1928) found (upper curve) that maximal bacteriocidal activity was produced by light of about 260 nanometers. This peak corresponded to the ultraviolet absorption peak of nucleic acid constituents (A, lower curves) rather than those of protein (curve B).

mice. Griffith found that if he heat-killed a culture of the
virulent (smooth) strain and inoculated the cells into a mouse,
no effect would be noted, but if the dead cells were
accompanied by a number of live, avirulent (rough) cells, the
mouse would die (Fig. 6). From the heart blood of the mouse
Griffith was able to isolate viable pneumococci exhibiting the

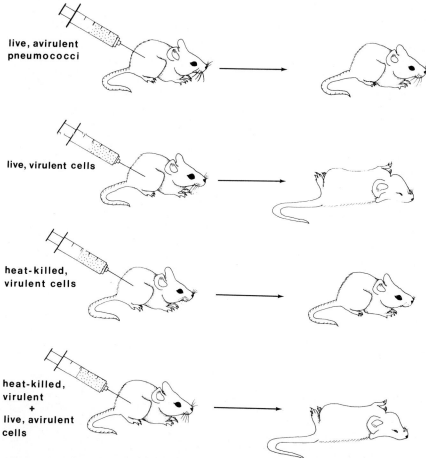

Figure 6. Griffith's discovery of transformation (1928). Neither live, avirulent
nor heat-killed virulent pneumococci had any effect on mice, but if the two
are mixed together and inoculated into a mouse (lowermost figure), the
mouse would succumb and from its heart blood could be isolated live, virulent
pneumococci.

polysaccharide capsule, smooth colonial morphology, and virulence. Evidently, the dead, smooth cells could still transfer information to the rough cells that enabled the latter to produce the capsule and thus acquire virulence. The factor that was transferred has been called *transforming principle,* its existence being demonstrated by Alloway in 1933, who managed to carry out Griffith's experiment using a cell-free extract of the smooth strain (Fig. 7).

The chemical nature of the transforming principle was worked out over a number of years by the American biochemist Avery and coworkers C. M. MacLeod and M. J. McCarty. After repeated purifications, it was found that the material appeared to be deoxyribonucleic acid. These results were so unexpected, so contrary to the then commonly accepted dogma that nucleic

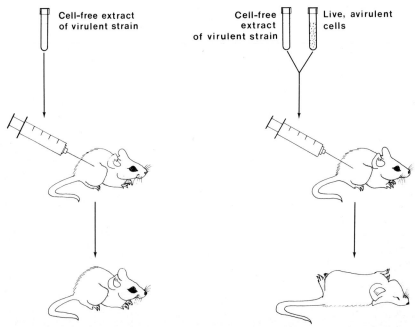

Figure 7. Transformation with cell-free extracts. Alloway (1933) incubated cell-free extracts of virulent pneumococci with live, avirulent bacteria. The resulting mixture was fatal to mice on inoculation (right side) even though the extract alone (left side) or avirulent cells had no effect.

acids were mere inert, structural molecules, that Avery hesitated publishing their results for almost a year. Finally, in 1944, the work appeared in print and as expected was accompanied by skepticism and doubt. The doubt actually had some substance, for Avery had never been able to purify completely the transforming principle, and there remained in it traces of protein. Avery's evidence showed that of all the enzymes tried, including ribonucleases, deoxyribonucleases, lipases, and proteases, only the deoxyribonuclease destroyed the activity of the transforming principle. In spite of these results many were not convinced that DNA could be capable of carrying genetic information. Others believed the active principle to be DNA, not as a genetic messenger, but as an agent that induced the change in the rough cells in an indirect manner. The controversy continued into the 1950's until evidence from another quarter firmly demonstrated the ability of DNA to act as a depository of genetic information.

Bacterial viruses, or bacteriophages, are obligate intracellular parasites of bacteria. Chemically they are roughly 50 percent protein and 50 percent nucleic acid. They attack susceptible bacteria by attaching to the external surface of the cell. In time the host bacterium lyses and releases hundreds to tens of thousands of virus particles identical to the one that initiated the infection. Hershey and Chase, in 1952, reported on an experiment involving phage and bacteria which elucidated the mechanism of phage infection and at the same time demonstrated the genetic role of DNA (Fig. 8). They grew one culture of bacteria infected with phage in a medium containing ^{32}P phosphate, and another infected culture in a medium containing ^{35}S amino acids. Thus, in the former case, the phage that developed would have their nucleic acid labeled with radioactive ^{32}P, since protein has no phosphorus, and conversely, in the latter case, the developing phage would have only their protein labeled with the radioactive ^{35}S. The radioactive phage were harvested and then allowed to infect nonradioactive bacteria in a nonradioactive medium, and soon after the initiation of the infection, the bacterial suspension was violently agitated to remove the attached phage particles from the cells.

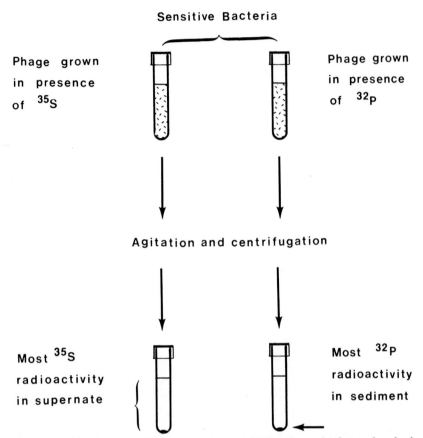

Figure 8. The Hershey-Chase experiment (1952). Bacteriophage that had been propagated in the presence of radioactive sulfur or phosphorus were mixed with sensitive host bacteria, following which the suspensions were vigorously agitated to remove adsorbed viral particles. The tubes were centrifuged sufficiently to sediment the bacterial cells, and the distribution of radioactivity was determined.

The suspensions were centrifuged to sediment the bacteria only, and the pellet and supernatant liquid were examined for radioactivity. Most of the ^{35}S was found in the supernatant liquid, while most of the ^{32}P was found in the bacterial sediment.

  Since it was assumed that whatever was necessary to propagate the bacteriophage would be found in the host cells, and the phage nucleic acid was found almost exclusively in them, then the conclusion was that the nucleic acid was the carrier of the genetic information necessary to produce the hundreds of progeny of the infecting particle.

THE STRUCTURE OF DNA

Much information concerning the primary structure of the nucleic acids was known by the 1920's. Specifically, it was demonstrated that the nucleic acids consisted of a pentose-phosphate backbone to which was attached the purine and pyrimidine bases (Fig. 9). The exact points of attachment of the phosphate groups on the sugar molecules, namely the 3' and 5' carbon atoms, were not verified until 1952, however. The unit building blocks of nucleic acids are the nucleotides, phosphate-pentose-base complexes that polymerize to form great chains with molecular weights of several million. Early chemical analyses of deoxyribonucleic acid extracted from several sources indicated that the bases appeared to occur in short chains of perhaps four nucleotides in equimolar amounts. From this work emerged the tetranucleotide hypothesis, suggesting that all DNA's are monotonously alike, regardless of source.

The technique known as *paper chromatography* was devised in the 1940's to separate complex mixtures of amino acids. The method was soon applied to other classes of compounds, and Erwin Chargaff used it to analyze acid hydrolysates of DNA. The technique enabled him to obtain quite accurate estimations of the base compositions of a number of different DNA's, and in 1950 he reported that all DNA's were not alike but had compositions characteristic of their source. DNA from mycobacteria exhibited a molar composition rich in guanine and cytosine, whereas DNA from members of the genus *Bacillus* was rich in adenine and thymine (Table I). Base compositions of nucleic acids are usually expressed in terms of the molar proportion of guanine and cytosine. Thus, mycobacterial DNA is said to have a composition of 72%G+C

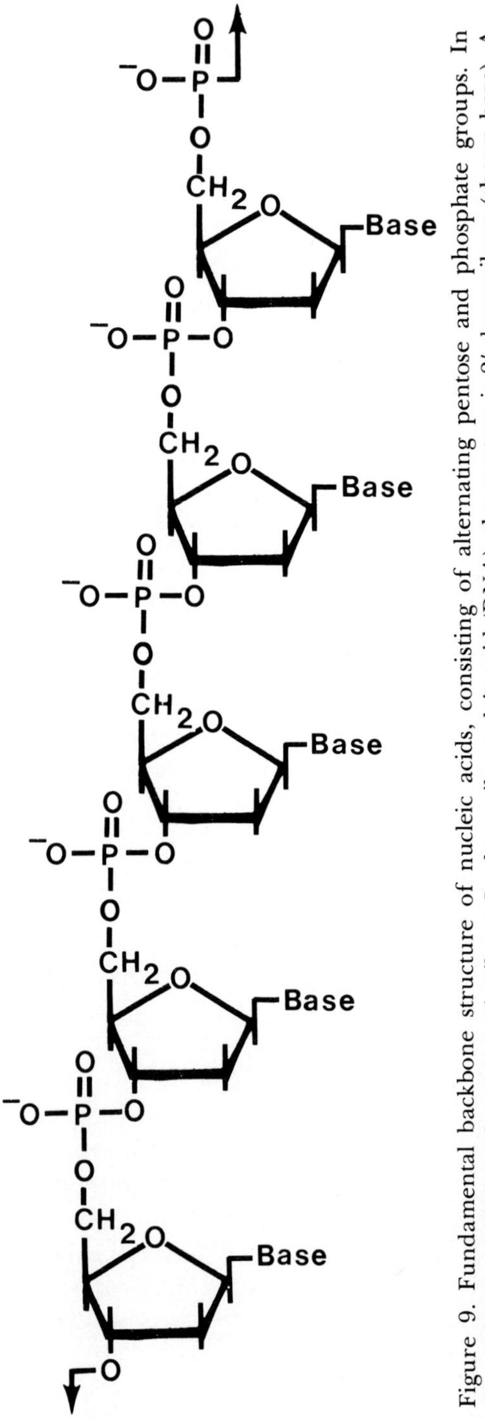

Figure 9. Fundamental backbone structure of nucleic acids, consisting of alternating pentose and phosphate groups. In ribonucleic acid (RNA), the pentose is ribose. In deoxyribonucleic acid (DNA), the pentose is 2'-deoxyribose (shown here). A purine or pyrimidine base is attached to the number 1 carbon of the pentose to complete the nucleotide building block of nucleic acids.

TABLE I
DNA BASE COMPOSITIONS OF SOME BACTERIAL GENERA*

Genus	Mole-Percent Guanine + Cytosine											
	25	30	35	40	45	50	55	60	65	70	75	80
Rhodospirillum								######	##			
Chromatium						#####	#########					
Chlorobium								####				
Myxococcus										####		
Archangium										##		
Cystobacter										#		
Chondromyces										##		
Flexibacter			###########									
Leptothrix									#			
Caulobacter						#####						
Asticcacaulis								####				
Spirochaeta							################					
Treponema			#####									
Leptospira			####									
Spirillum		######										
Bdellovibrio		#####			########							
Campylobacter			###									
Pseudomonas							################					
Azotobacter								####				
Rhizobium							#######					
Halobacterium				(Minor component)				####	###			
Brucella							###					
Escherichia					##							
Salmonella						####						
Klebsiella							#####					
Serratia							#######					
Proteus				#####	##########	##						
Vibrio				#####								
Aeromonas							#######					
Photobacterium				####								
Flavobacterium								########				
Hemophilus				####								

TABLE I—*Continued*

Genus	Mole-Percent Guanine + Cytosine											
	25	30	35	40	45	50	55	60	65	70	75	80
Bacteroides				#################								
Selenomonas							#############					
Neisseria				####								
Acinetobacter				#########								
Veillonella				#####								
Nitrobacter								##				
Thiobacillus								##				
Micrococcus			###########									
Staphylococcus			##########									
Streptococcus			###################						###############			
Bacillus			################################									
Clostridium							###					
Lactobacillus			##########	##	######							
Corynebacterium						###################						
Arthrobacter								###############				
Mycobacterium									###############			
Rickettsia		##										
Mycoplasma	##	#	######################									

*Data from various sources. Positions of symbols may be approximate.

and this was arrived at by dividing the total moles of the four bases into the moles of guanine and cytosine, times 100:

$$\%G+C + \frac{\text{moles (G)}+\text{moles (C)}}{\text{moles (G)}+\text{moles(C)}+\text{moles(A)}+\text{moles(T)}} \times 100$$

Of greater significance was the observation of Chargaff that for a given DNA and allowing for some experimental error, the moles of adenine seemed to equal the moles of thymine, and the moles of guanine seemed to equal the moles of cytosine (Table II). Consequently, the moles of purines equaled the moles of pyrimidines.

TABLE II
PURINE AND PYRIMIDINE COMPOSITIONS OF VARIOUS DNA's

Source	Adenine	Thymine	Cytosine	Guanine
Bovine thymus*	28.2	27.8	21.2	21.5
Bovine spleen*	28.2	28.2	21.0	21.2
Salmon sperm	29.7	29.1	20.4	20.8
Human thymus	30.9	29.4	19.8	19.9
Escherichia coli	24.7	23.6	25.7	26.0
Mycobacterium tuberculosis	15.1	14.6	35.4	34.9
Sea urchin sperm	32.8	32.1	17.3	17.7

*Figures are in moles per gram-atom of phosphorus.
DNA from sources marked * contain small amounts (1 to 2%) 5-methylcytosine.

Also in 1950, an American geneticist, James Watson, was spending a postdoctoral year in Denmark trying to learn as much as he could about the biochemistry of DNA. Watson soon lost interest in the subject until he happened to hear a paper delivered by Maurice Wilkins. Wilkins was a British X-ray crystallographer who had carried out structure studies on DNA for a number of years. In order to study a material by X-ray crystallography, it must possess the repeating characteristics of a crystal, for the technique depends on the diffraction of a beam of X-rays as it passes through the crystal lattice. The patterns that the diffracted beam form can be used to deduce the secondary structure of the crystal (Fig. 10). What excited Watson was the realization that DNA possessed crystalline properties. Very soon after Wilkins' talk, Watson read Linus

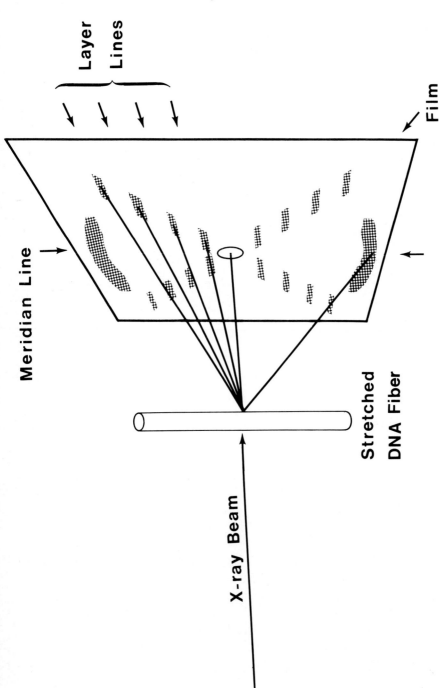

Figure 10. X-ray diffraction analysis of DNA. A collimated beam of x-radiation is aimed at a stretched fiber of DNA. The periodicity of the DNA structure causes the X-ray beam to be diffracted in a regular pattern, which can be visualized by placing a sheet of photographic film behind the DNA. The lack of reflections along the meridian line except at its extremities denote a helical structure, and the angle the layer lines make with the meridian line, and their arrangements and intensities allow one to calculate the number of turns in the helix, and its dimensions.

Pauling's paper on the determination of protein structure by X-ray crystallography.

The path was clear. Watson managed to have his postdoctoral fellowship transferred to Cambridge University and the laboratory of another X-ray crystallographer, Max Perutz. It was here that Watson met Francis Crick, an English physicist working toward his doctoral degree. The two in a little over a year accumulated the small amount of information available on DNA, including that of Chargaff and of Wilkins and his collaborators, and devised a model of the DNA molecule. The model consists of the following: two chains of nucleotides interwound about one another to form a double helix, with the purine and pyrimidine bases facing inward and forming specific, hydrogen-bonded pairs. The model fitted the information at hand. The X-ray work showed the DNA to be a helical molecule with a diameter of about 20Å. Strong reflections appeared at 34Å intervals, and again at 3.4Å. The 34Å intervals appeared to be the repeated twists of the helix, the 3.4Å as the individual base pairs along the helix (Fig. 11). The manner by which the inward facing bases paired up had been suggested by the results of Chargaff. Wherever there was a guanine on one strand of the helix, a cytosine was found on the opposite chain. Wherever there was a thymine, opposite was an adenine.

The attractive force that held the bases together, and ultimately the two strands of the molecule, was the specific hydrogen bonding that was most favored between guanine and cytosine, and between thymine and adenine (Fig. 12). Watson and Crick published the results of their model in the April 25th, 1953, issue of the British journal *Nature*. The paper started out, "We wish to suggest a structure for the salt of deoxyribose nucleic acid (DNA). This structure has novel features which are of considerable biological interest." Watson and Crick received the Nobel Prize in 1962 for their suggestion.

Certain aspects of the Watson and Crick structure should be pointed out here. Notice in Figure 13 that the opposite strands of the DNA are not mirror images of one another, for if you read off the carbon-phosphate diester bonds of the backbone

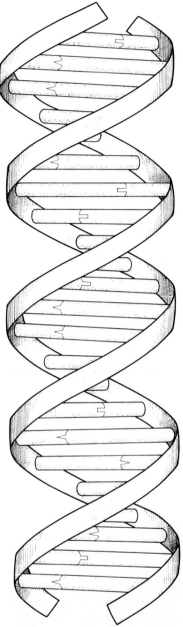

Figure 11. The Watson-Crick model of DNA. Two pentose-phosphate backbone structures are wound about one another in helical fashion. The strands are held together through specific hydrogen-bonded base pairings: adenine with thymine (rectangular joinings), and guanine with cytosine (triangular joinings).

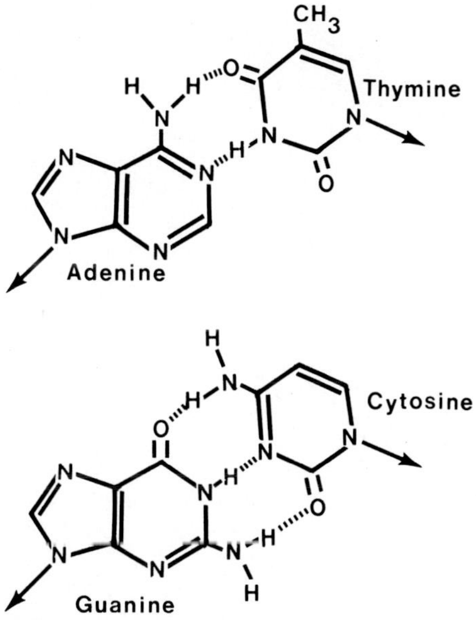

Figure 12. Hydrogen bonding between purine and pyrimidine bases. The arrangement of nitrogen and oxygen atoms in adenine and thymine (upper figure) favor these bases sharing two hydrogen bonds. Hydrogen bonding is also favored between a guanine molecule and a cytosine molecule. The specificity of such bonding (adenine with thymine, guanine with cytosine) is the basis for the Watson-Crick structural model of DNA as well as the model for its replication.

down one strand, they are 3'-5', 3'-5', 3'-5', etc. But reading down (in the same direction) the other strand, the bonds are 5'-3', 5'-3', 5'-3'. Such a structure is known as being antiparallel, and this feature will have deep significance when we cover the enzymology of DNA replication in later sections.

One of the characteristics of DNA is its fragility, and unless great care is taken in its handling, a DNA molecule will fragment into many pieces. Much of this fragility is due to the double strandedness of the molecule, giving it a rigidity and a

34

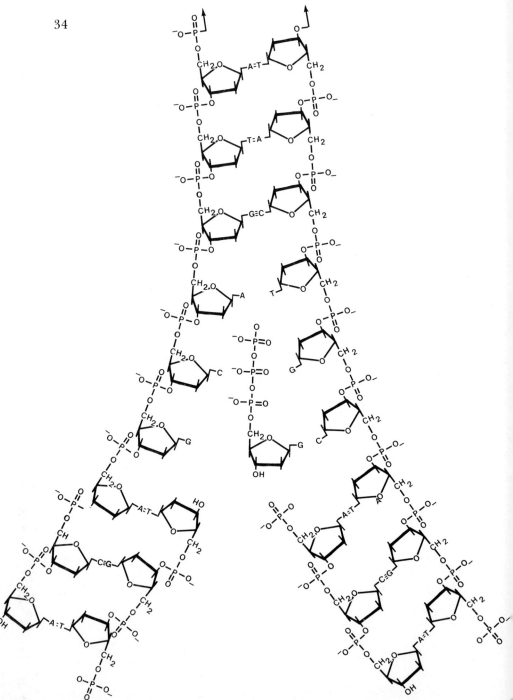

Figure 13. Detail of DNA structure and its replication. Nucleoside triphosphates are arranged along opposite strands via specific base pairing. Nucleotides are enzymatically joined through phosphodiester bonds, with the release of pyrophosphate.

susceptibility to breakage. Only recently have methods been devised to extract whole DNA molecules from microbial cells. For this reason much of the early data (pre-1940 particularly) in the literature concerning the molecular weights of DNA from various sources were in error.

The molecular weights of some representative DNA molecules are shown in Table III. It has been established that *E. coli* DNA for example is represented by a single molecule of molecular weight 3×10^9, representing about 3×10^6 nucleotide base pairs. The total length of such a molecule, if stretched out, would be 1100 μm.

TABLE III
MOLECULAR WEIGHTS OF SOME DNA's*

Source	Molecular weight	Number of chromosomes per cell
Polyoma virus	4×10^6	1
T2 phage	1.3×10^8	1
E. coli	3×10^9	1
Aspergillus	3×10^{10}[†]	8
Drosophila	5×10^{10}[†]	4
Man	1.6×10^{12}[†]	23
Corn	4.5×10^{12}[†]	10

*Compiled by Silver, S.: Molecular genetics of bacteria and bacteriophage, *Prog Biophys Molec Biol, 16:*193, 1966. Reprinted by permission of Pergamon Press Ltd.
[†]Aggregate molecular weights

DNA AND CHROMOSOMES

You recall that in Chapter One we referred to Walter Sutton's proposal that the intensely staining bodies known as chromosomes were the depositories of genetic information of the cell. Cytochemists have shown us that chromosomes of higher plants and animals are made up of roughly equal proportions of protein and DNA, and in this chapter we just established that it is the DNA that is the actual informational molecule. The chromosomes are found in the nucleus of the cell, the number of chromosomes making up the cell's complement depending upon the species. As we examine the nuclei of cells from more complex organisms down to simpler forms such as amoebae or fungi, we find the same general structure. The nuclei are well defined and surrounded by a membrane. The chromosomes are also well defined and undergo the characteristic process of mitosis during certain phases of the cells' cycle. But as we pass on to the bacteria, we find a different state of the nucleus. We find no surrounding membrane and no obvious mitotic apparatus. The bacteria appear to have but one chromosome, consisting of a single molecule of DNA. Cells with such simpler nuclear structures are referred to as being *procaryotic,* whereas those cells that exhibit nuclear membranes and mitotic apparatuses are called *eucaryotic.* Besides bacteria, examples of other procaryotic organisms are the actinomycetes and the mycoplasmas.

THE *IN VIVO* REPLICATION OF DNA

The answer to how a DNA molecule may replicate itself also presented itself to Watson and Crick soon after the establishment of their model. The helix could unwind, exposing unpaired bases that would act as specific templates for the formation of new strands (Fig. 13). Each guanine and thymine would attract a cytosine or adenine, respectively, and vice-versa.

In 1963 John Cairns devised a technique for the visualization of a whole bacterial chromosome. Cairns grew a culture of *E. coli* in the presence of tritiated thymidine so that its DNA was labelled with the radioactive isotope of hydrogen. The cells were then placed on microscope slides and gently lysed with lysozyme. The contents of the cells were allowed to spread out onto the slide undisturbed. The slides were dried and in the dark a photographic emulsion was poured over the cell debris. The emulsion was similar to the layer of light-sensitive gel coated onto celluloid to make ordinary photographic film. The emulsion hardened, and the slides were stored in the dark for a few months. The emulsion layers were then developed and examined under a microscope. The tritium, being radioactive, emits beta particles on disintegration, and wherever a beta particle struck the emulsion, a blackened grain would result. Thus, an image of the DNA molecule was produced in the photographic emulsion. Figure 14 shows one of Cairns' autoradiograms, as they are called. The first feature that strikes one when seeing this figure is that the image of the DNA is circular, with no apparent ends. There had been genetic evidence for a circular bacterial chromosome, but this had been the first physical evidence of it.

In addition, this chromosome apparently has been caught in the midst of replication, with its daughter chromosome still attached. Figure 15 is a series of drawings of the steps that led to the configuration in Cairns' autoradiogram. While the figure shows but a single replicating fork, there is evidence that under certain circumstances both forks may be involved in replication, in which case replication is bidirectional.

When a DNA molecule replicates, at least two or three outcomes can result relative to the distribution of the "old" and "new" strands. When the daughter strands have completed replication, they can segregate in a manner known as *conservative replication* in which one of the double-stranded molecules consists of all new material and the other consists of all old material. In *semiconservative replication,* each of the daughter molecules consists of one new strand and one old strand. Other modes such as one combining both types of replication and known as *dispersive replication* are also possible (Fig. 16).

One of the first attempts to determine which of these modes operates in bacteria was carried out by Meselson and Stahl (1958). A new technique had just been devised for the separation of macromolecules according to density. The technique involved the preparation of concentrated solutions of certain salts, such as cesium chloride. If such a solution were placed in a tube in the ultracentrifuge and subjected to gravitational forces approaching 100,000 x g, the heavy cesium ions will begin to move towards the bottom of the tube. Eventually a near linear concentration gradient is formed through the length of the tube, with the higher concentrations being toward the bottom of the tube. The relatively great density of such solutions thus results in a very perceptible density gradient throughout the tube, as well. If particles or

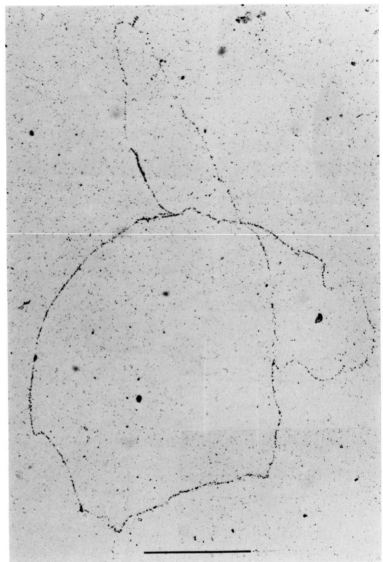

Figure 14. Autoradiogram of replicating DNA. An image of an entire bacterial chromosome has been captured by autoradiography. Bacterial DNA, made radioactive by incorporating tritiated thymine during growth of the bacteria, is spread over a glass slide and covered with a photographic emulsion. During extended storage, beta particles emitted from the radioactive thymine ultimately result in the blackened grains that make up the image.
(From Cairns, J.: *Cold Spring Harbor Symp Quant Biol, 28:*43, 1963)

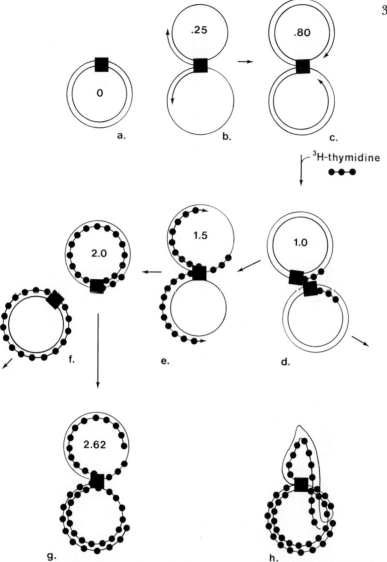

Figure 15. Diagrammatic presentation of the sequence of events leading to the configuration of the chromosome shown in Figure 14. For clarity, the double-stranded bacterial chromosome (a.) is shown "butterflied" (b. through h.). Numerals depict generation numbers. After about 0.80 generations (c.), the cells were transferred to a medium containing tritiated thymidine, resulting in radioactive strands of DNA made after the transfer. At generation 2.62 (g.), the cells were spread onto glass slides, gently lysed, and covered with a photographic emulsion. Figure g. is redrawn in figure h. to simulate more closely the configuration assumed by the chromosome in Cairn's autoradiogram (Fig. 14).

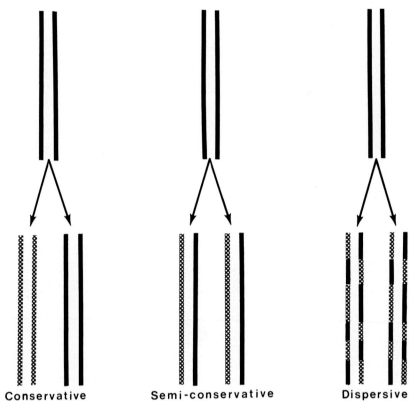

Conservative   **Semi-conservative**   **Dispersive**

Figure 16. Modes of DNA replication. Shown here are three possible modes of DNA replication. CONSERVATIVE: double-stranded parent molecules replicate, forming one molecule composed of all new material, and one molecule conserving the parental molecule. SEMICONSERVATIVE: replication may produce two molecules each of which is composed of one new strand and one old (parental) strand. DISPERSIVE: Progeny molecules may be composed of mixtures of new and old material.

macromolecules also happen to be in the tube, they too would come under the influence of the gravitational force and move towards the bottom of the tube. As the particles pass through the density gradient, which increases as they approach the bottom, when they reach a point in the gradient that corresponds to their own density, they would cease to sediment. Particles that happen to be below the point in the tube that corresponds to their density will float against the forces of the gravitational field until they reach the appropriate density band. Thus a mixture of particles could be separated according to their densities by such a technique.

Meselson and Stahl grew *E. coli* for fourteen generations in the presence of $^{15}NH_4Cl$. ^{15}N is a nonradioactive isotope of nitrogen, and its presence in a molecule imparts additional density to it. Thus, all of the nitrogen-containing molecules of the bacteria exhibited an increase in density.

Of particular interest was the DNA of the cells, which had acquired an increase in density of 0.014 g/cm^3. Such a small difference was easily distinguished by the density gradient centrifugation technique. The ^{15}N-labelled cells were then transferred to a medium containing ^{14}N nitrogen sources and incubated while aliquots were removed at frequent intervals. The cells from each sample were lysed, and the lysates were subjected to the density gradient centrifugation. After one generation in the ^{14}N medium, the DNA of the cells was seen to shift its position relative to the initial ^{15}N DNA. After two generations, an additional shift was observed, this time to a position corresponding to the density of normal ^{14}N DNA. The DNA isolated from the cells after one generation appeared to locate on the density gradient in a position exactly half way between the ^{15}N DNA and the ^{14}N DNA. This is what one would expect if the DNA were undergoing semiconservative replication during that generation in the ^{14}N medium (Fig. 17). As a control experiment, Meselson and Stahl separated the DNA strands of intermediate density by heating and determined the densities of the single strands. The densities corresponded to those of a totally ^{14}N DNA and a totally ^{15}N DNA. Thus, the possibility of a dispersive mode of replication was eliminated, and the semiconservative mode was supported.

Referring back to the autoradiogram of Cairns (Fig. 14), this also confirmed the existence of the semiconservative mode of bacterial DNA replication. If one counts the blackened grains along the image of the DNA, it is found that there are exactly twice as many grains along one portion as there are along another. The double-grained region represents a portion of the DNA molecule that had replicated for part of a second

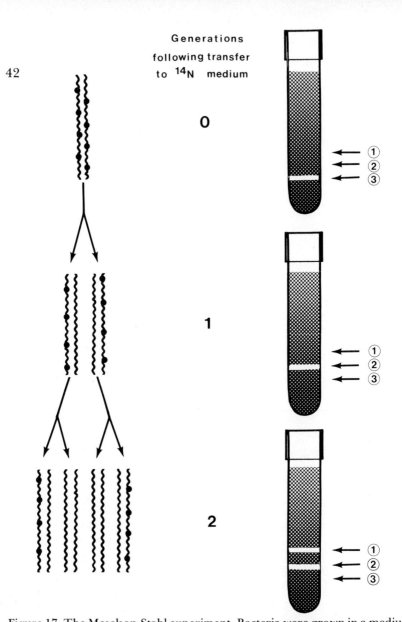

Generations
following transfer
to ¹⁴N medium

0

1

2

Figure 17. The Meselson-Stahl experiment. Bacteria were grown in a medium containing ¹⁵N nutrients, resulting in cells with DNA of greater than normal density, as demonstrated by density gradient centrifugation (right side of figure). In the uppermost density gradient tube, the ¹⁵N DNA bands at position 3, whereas normal DNA would band at position 1. Then after one generation in the presence of ¹⁴N nutrients, the DNA thus produced bands at position 2, a point that represents a density intermediate to normal and ¹⁵N DNA. After two generations in the ¹⁴N medium, extracted DNA forms two bands, one at the intermediate position, and one at the normal position. The pattern of the bands is most easily explained if the DNA undergoes semiconservative replication in which (left side) the uniformly dense ¹⁵N strands separate on replication to form molecules of intermediate density at generation 1, and two populations of normal and intermediate densities at generation 2.

generation in the presence of the radioactive thymidine. If conservative replication were operating, part of the DNA would be labelled, but with uniform grain distribution, and part would be unlabelled, and not appear on the autoradiogram. With dispersive replication, double-labelled and single-labelled portions would appear in a random fashion on the autoradiogram.

The Cairns model as described above and in Figure 15 is one of two fundamental mechanisms of DNA replication that are currently favored. The second mechanism, known as the Rolling Circle model, was presented by Gilbert and Dressler in 1968. This model may occur in the replication of certain bacteriophages and in the transfer of bacterial DNA during conjugation. The model envisions the breakage of one of the nucleotide strands and the unrolling of the second, closed-loop strand as replication proceeds (Fig. 18). Part of the model

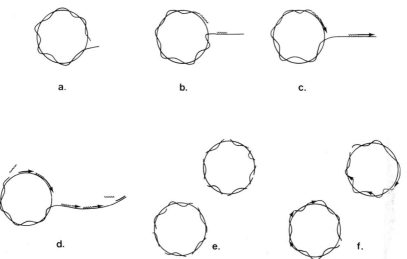

Figure 18. The Rolling Circle model for DNA replication, combining features proposed by Gilbert and Dressler, Okazaki, and others. One strand is nicked (a.), allowing it to unwind. Short segments of RNA (b.) are formed on the separated strands, which act as primers for the formation of short fragments of DNA (c.). Following release of the RNA primers (d.), the gaps left by the RNA (e.) are filled by repair enzymes (f.). Arrows denote the direction of replication.

includes the attachment of the loose end of the open strand to some anchoring site, perhaps on the cell membrane. One feature of the Rolling Circle model is that it explains how linear multiple copies, or *concatemers*, of DNA are possible, as has been observed in some phages. Replication need not cease at the starting point on the closed loop, but can continue for an additional turn or fraction of a turn, resulting in the replication of more than one genome's length of DNA. Once the closed loop is released from the open strand, the latter can reform a closed configuration through a recombination process between homologous sequences of nucleotides at the ends of the strand.

While the two models cited here differ in many respects, they appear to share certain generalized steps as hypothesized by Sugino and Okazaki and others:

1. One of the parental strands unwinds (Fig. 18a), exposing unpaired nucleotide bases.
2. A short (50 nucleotides) segment of RNA is formed at the nicked end of each parental DNA strand, forming an RNA-DNA double-stranded hybrid segment (Fig. 18b).
3. Using the 3′ end of the RNA segment as a primer, DNA synthesis is initiated and proceeds along the parental strands always in a 5′——→3′ direction until the end of the unwound segment is reached (Fig. 18c).
4. The primer RNA segment is released, leaving a single-stranded gap (Fig. 18d).
5. Nicks formed in step 1 are rejoined (Fig. 18e).
6. The gaps left by the RNA primer are filled using the opposite strand as a template (Fig. 18f).

Each step in the proposed model for DNA replication is supported by evidence pointing to the existence of one or more enzymes known to carry out that step:

1. An enzyme has been found in normal and bacteriophage-infected *E. coli* that appears to catalyze the unwinding of a twisted DNA molecule.
2. Okazaki was the first to demonstrate the presence of short segments of DNA during DNA replication (so-called *Okazaki fragments*). These fragments are frequently found to be covalently attached to small segments of RNA at the

RNA 3′ end. Furthermore, the drug rifampicin, known to specifically block RNA ploymerase action, also blocks DNA replication.

3. The DNA polymerases so far studied appear to require the 3′-hydroxy end of a primer nucleotide fragment on which to initiate chain elongation.
4. An enzyme known as ribonuclease H has been found in eucaryotic cells. One of the actions of RNase H is to release RNA that is covalently bound to DNA at the RNA 3′ terminus. Similar enzymes may be found in bacteria.
5. DNA ligase, an enzyme that can be isolated from normal or phage-infected bacteria, can join adjacent polynucleotides together.
6. Bacteria and other cells possess several enzymes capable of filling nucleotide gaps that may occur in one of the strands of a double-stranded DNA molecule.

The depictions of the bacterial chromosome in Figures 15 and 18 probably are misleading in that they imply a loose, open configuration. Electron micrographs and other evidence suggest the bacterial chromosome to be in actuality a dense, tightly wound skein of superhelical folds, probably held together by short links of RNA. Evidence also points to the presence of at least one single-strand break within each fold of the DNA to permit unwinding during replication.

THE *IN VITRO* REPLICATION OF DNA

A major step in the understanding of the enzymology of DNA replication came in 1957 when Arthur Kornberg reported on the successful replication of DNA *in vitro* using a polymerase extracted from cells of *E. coli*. The components of Kornberg's system are listed in Table IV. If any of the constituents were omitted from the reaction mixture, no polymerization was observed. The DNA that was formed with the complete system had a base composition very nearly identical to that of the template DNA (Table V). Careful analysis of the polymerization reaction products revealed that

TABLE IV
COMPONENTS OF KORNBERG'S REACTION FOR THE *IN VITRO* SYNTHESIS
OF DNA

Deoxythymidinetriphosphate
Deoxycytidinetriphosphate
Deoxyguanosinetriphosphate
Deoxyadenosinetriphosphate
Template DNA
Mg^{++}
DNA polymerase

TABLE V
BASE COMPOSITIONS OF SYNTHETIC DNA MADE FROM VARIOUS
NATURAL TEMPLATES IN THE KORNBERG *IN VITRO* EXPERIMENTS*

Source	Adenine	Thymine	Cytosine	Guanine
Bovine thymus				
Template	28.5	26.3	21.3	22.5
Synthetic	29.8	29.8	20.8	20.3
E. coli				
Template	25.0	24.3	26.3	24.5
Synthetic	26.0	25.0	24.5	24.3
Mycobacterium phlei				
Template	16.3	16.5	33.8	33.5
Synthetic	16.5	20.0	29.3	33.5
Bacteriophage T2				
Template	32.8	33.0	15.5[†]	16.8
Synthetic	33.2	32.2	15.5	17.3

*From Lehman, I.R.: *Ann NY Acad Sci, 81:*745, 1959
[†]hydroxymethylcytosine; figures are in mole-percent.

the polymerase was only capable of adding nucleotide triphosphates to the 3′ hydroxy end of the existing chain. Kornberg was awarded the Nobel Prize for this work in 1959, and the techniques established by him have subsequently made it possible to dissect many of the steps in DNA replication.

The polymerase first isolated by Kornberg is now referred to as DNA polymerase I, the role of which *in vivo* is not completely known, but appears to be one of filling gaps left in the DNA strands when replication is near completion. Other proteins with DNA polymerase activity have been isolated, such as DNA polymerase II, which appears to be a repair enzyme and not directly associated with DNA replication. A third enzyme, DNA polymerase III, can act alone or can form multiunit molecules with other proteins to catalyze the elongation of DNA chains. It appears to play a central role in the replication of DNA.

Soon after Kornberg had demonstrated the *in vitro* synthesis of DNA, the question was raised whether the synthetic DNA retained the biological activity of the template molecule from which it was formed. Attempts to show such activity in synthetic DNA failed until Goulian, Kornberg and Sinsheimer (1967) managed to produce synthetic bacteriophage DNA that did possess biological activity. As template for *in vitro* synthesis they chose the single-stranded DNA of bacteriophage φX174. A double stranded molecule thus formed was subjected to mild nuclease action, the effect of which was the release of intact, single-stranded synthetic molecules of DNA. These were actually complementary strands of the original phage DNA strands and were thus referred to as "−" strands. The − strands were tested in a transfection experiment (see Chapter Six) and were shown to possess all of the biological activity necessary to initiate and complete the infection of bacterial cells. If the synthetic − strands were used as templates to produce totally synthetic double-stranded phage DNA, it too possessed biological activity, as did the synthetic + strands separated from the double strands. We therefore see that aside from the uncertainty of the role played by the Kornberg polymerase in *in vivo* DNA replication, it is nevertheless capable of catalyzing the replication of DNA molecules with sufficient fidelity to retain their biological activity.

DETERMINATION OF DNA COMPOSITION

Since Chargaff's investigations a number of other techniques have been devised to determine the base composition of nucleic acids. The two most widely used methods are based on physical parameters of the DNA, and while requiring somewhat more sophisticated equipment than Chargaff's paper chromotography, are now the methods of choice. One technique involves the heating of a solution of the DNA until the strands separate. If one observes the heating in a specially equipped ultraviolet spectrophotometer (Fig. 19), a temperature is reached where the absorption suddenly rises almost 50 percent (Fig. 20). The

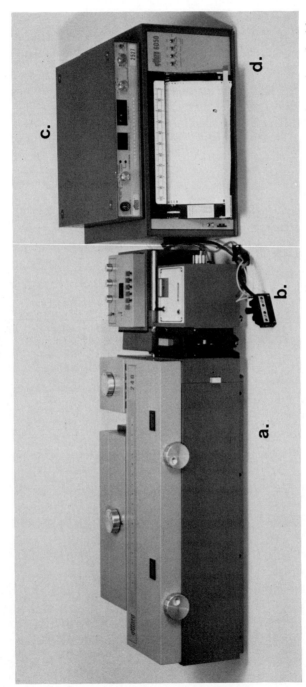

Figure 19. Apparatus for determining melting points of DNA. An ultraviolet spectrophotometer (a.) is equipped with a heated cuvet holder (b.). The rate of heating is precisely controlled by component c. A recorder (d.) plots optical density against temperature. (Photo courtesy Gilford Instrument Laboratories, Inc.)

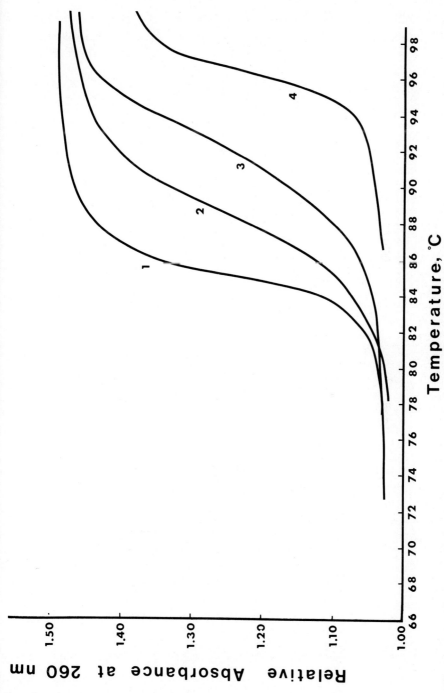

Figure 20. Melting points of some bacterial DNA's. 1. *Streptococcus pneumoniae*, 2. *E. coli*, 3. *Serratia marcescens*, 4. *Mycobacterium phlei*. (From Marmur, J., and Doty, P.: *Nature, 183*:1427, 1959).

midpoint at which this so-called *hyperchromic shift* occurs is known as the melting point or T_m of the DNA. It was observed that under standard conditions of pH and ionic strength of the solvent, the T_m for a given DNA is constant, but is proportional to its base composition (Fig. 21). To estimate the base compositon of a DNA by its T_m one must assume it is double

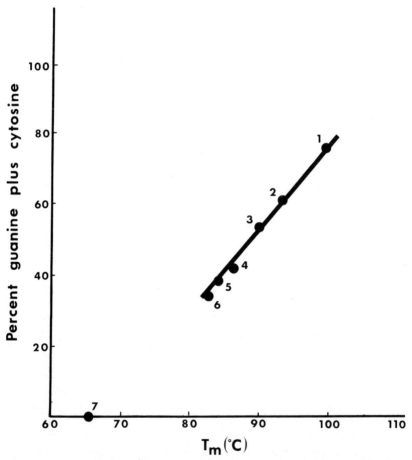

Figure 21. Relationship between melting point (T_m) of DNA and its base composition. 1. *Mycobacterium phlei.* 2. *Serratia marcescens.* 3. *E. coli.* 4. Calf thymus. 5. *Streptococcus pneumoniae.* 6. Bacteriophage T4r. 7. Poly AT (a synthetic polymer of random adenylates and thymidylate residues). (From Marmur, J., and Doty, P.: *Nature, 183:*1427, 1959)

stranded and consists of only adenine, thymine, cytosine, and guanine nucleotides. We shall meet a number of exceptions to these assumptions in later chapters.

The relationship between %GC and melting point is due in part to the three hydrogen bonds that hold guanine-cytosine base pairs together, while there are only two for adenine-thymine pairs. The richer a DNA is in GC pairs, the more hydrogen bonds are holding the strands together and presumably the more energy in the form of heat is required to bring about strand separation, or *helix-coil transition,* as it is frequently referred to.

The hyperchromic phenomenon itself appears to be due to the difference in the distribution of the electrons associated with the purine and pyrimidine bases in the regular double stranded helix compared with the random, single stranded coil. Ultraviolet light is more strongly absorbed in the latter case than in the former.

Density can also be used as a means of estimating the base composition of a DNA. The most widely used means of determining the density of a DNA is by the density gradient centrifugation technique discussed earlier. By means of standard DNA's and careful technique, it is possible to determine the density of a DNA to 0.001 g/cc. The base composition is estimated by applying the formula:

$$\%GC = \frac{\rho - (1.660)}{(0.00098)}$$

Where ρ=density of DNA (Fig. 22)

HYBRIDIZATION OF NUCLEIC ACIDS

The thermal denaturation of double stranded DNA can be at least partially reversed by allowing the melted DNA to cool to about 65°C and holding it at that temperature for several minutes. Such renaturation can be almost 100 percent complete with viral DNA, while larger DNA molecules from bacteria achieve no more than about 50 percent renaturation. Molecular weight determinations of renatured DNA show that the

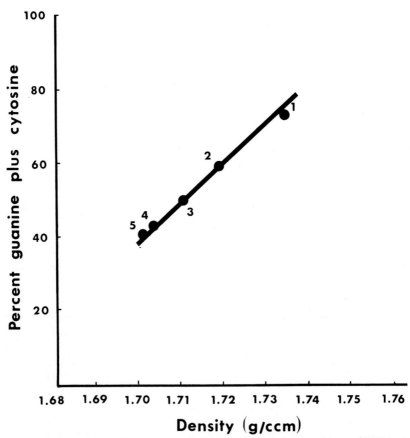

Figure 22. Relationship between density and base composition of DNA. 1. *Myco. phlei*, 2. *Serratia marcescens*, 3. *E. coli*, 4. Salmon sperm, 5. *Streptococcus pneumoniae*. (From Sueoka, H., Marmur, J., and Doty, P.: *Nature, 183:*1429, 1959).

reaction involves the actual reestablishment of the complementary, double-stranded state.

That complementarity is involved in DNA renaturation has been shown by mixing DNA's from two sources and subjecting the mixture to denaturation and renaturation. Single strands from source A will renature, or hybridize, with strands from source B to the degree of their taxonomic relatedness. Apparently, hybridization is only possible when at least some regions of homologous base sequences are localized on the two strands.

Hybridization is also possible between strands of DNA and homologous strands of RNA, or between RNA's from different sources. This technique has proven to be one of the most valuable tools to be developed in many years. It has not only been of some value to taxonomists, but it has proven to be extremely useful in the isolation and purification of specific regions of the DNA of various organisms.

TABLE VI

NUCLEIC ACID HYBRIDIZATION AMONG MEMBERS OF THE GENUS
*LACTOBACILLUS**

Source of DNA	Extent of hybridization with RNA made from L. leichmannii ATCC 7830 DNA
Lactobacillus leichmannii ATCC 7830	100%
L. leichmannii ATCC 4797	100%
L. delbrückii ATCC 9649	34.4%
L. lactis 39-A	98.4%
L. acidophilus Farr	100%
L. acidophilus IFO 3532	7.4%
Sporolactobacillus inulinus	0
Escherichia coli	0

DNA was extracted from ATCC strain 7830 of *L. leichmannii* and was used as a template for the making of complementary RNA. The RNA was then tested for its degree of hybridization with DNA extracted from various species of lactobacilli and other organisms.

*From Miller, A., Sandine, W. E., and Elliker, P. R.: *Canad J Microbiol, 17:*625, 1971. Reproduced by permission of the National Research Council of Canada.

Table VI shows the results of some hybridization experiments in which DNA from various members of the genus *Lactobacillus* was tested for extent of hybridization with RNA produced from a DNA template of a strain of *L. leichmanii*. It would appear that *L. lactis* is more closely related to *L. leichmanii* than is *L. delbrückii,* for example. It had long been observed by the authors of this work that the Farr strain of *L. acidophilus* demonstrated the same fermentation reactions and similar DNA base composition as those of *L. lactis,* and it was suggested that this organism may in fact belong in the species *lactis.* Certainly the hybridization results support this hypothesis.

Chapter Three ―――――――――――――――――――

PROTEIN SYNTHESIS

―――――――――――――――――――――――――――――

THE GENETIC CODE

SOMETHING LIKE 80 to 90 percent of the carbon assimilated by a bacterial cell and a like proportion of the energy consumed by it are funneled into the synthesis of protein molecules. The proteins, principally in the form of enzymes, consist of linear polymers of as many as several hundred amino acids, and the kind, number, and specific sequence of the amino acids in a given protein are what ultimately rule its activity. Information on amino acid sequences has been determined for a number of proteins. For example, the structure of pancreatic ribonuclease is shown in Figure 23. It consists of a single chain of 124 amino acids, beginning with lysine at the amino-terminal end, and terminating with valine at the carboxy end. Four disulfide bridges bind the chain into a relatively compact shape.

As Beadle and Tatum set forth in 1941, for each enzyme produced by a cell, there is a specific genetic region responsible for the formation of that enzyme. We have seen in Chapter Two that genetic material consists of deoxyribonucleic acid, linear polymers of the four deoxyribonucleotides. We are now confronted with the question of how the information presumably encoded in the DNA is translated into the primary structure of a given protein. It would appear that a sequence of purine and pyrimidine bases could spell out an amino acid, the next sequence, the next amino acid, and so on, until the hundred or so amino acids of the protein are designated. A bacterial chromosome contains about five million base pairs,

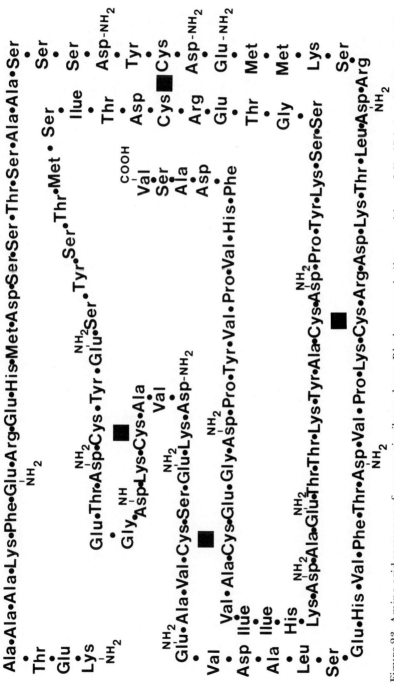

Figure 23. Amino acid sequence of pancreatic ribonuclease. Black squares indicate positions of disulfide bonds. (From Smyth, D. G., Stein, W. H., Moore, S.: *J Biol Chem, 238*:277, 1963).

certainly enough capacity to encode the primary structures of the maximum of 4000 different proteins bacterial cells are purportedly capable of producing, even if it takes eight to ten bases to designate each amino acid.

An examination of a three-dimensional model of DNA prompted a suggestion by physicist George Gamow in the early 1950's that the DNA could act as a direct template for the formation of protein molecules. The grooves resulting from the twisted, double stranded configuration could accomodate the amino acids, the particular shape of the spaces in the groove being governed by the adjacent base pairs and thus, specific for given amino acids. However, for a number of reasons the proposal was deemed unsatisfactory. Basically, it was difficult to visualize sufficient template sites on the relatively simple DNA chain that would accomodate the variety of side groups of the twenty naturally occurring amino acids. What was known of the physical dimensions of proteins and nucleic acids (Fig. 24), among other things, prompted Crick, for example, to remark that he did not think that anybody looking at DNA or RNA would think of them as templates for amino acids. Furthermore, it had just been agreed that protein synthesis was carried out at sites remote from the cell's DNA, on cytoplasmic particles known as ribosomes. There must be a messenger molecule that carried the information from the DNA to the ribosomes.

Ribosomes are ellipsoidal particles some 200 Å in their maximum dimension and consist of about 35 percent protein and 65 percent ribonucleic acid (Fig. 25). It was suggested that it was the ribosomal RNA that was the messenger molecule. The ribosomal RNA (rRNA) could replicate off of the DNA template in the same manner that Watson and Crick had suggested DNA replicates. Since a rapidly growing bacterial cell has about 10,000 ribosomes in its cytoplasm, it was concluded that each ribosome could be responsible for the formation of one of the 4000 different proteins the cell may need to produce. This period in the late 1950's was therefore referred to as the "one gene-one ribosome-one enzyme" era.

Upon careful analysis, ribosomal RNA from several species appeared quite similar in base composition, some 50 to 55

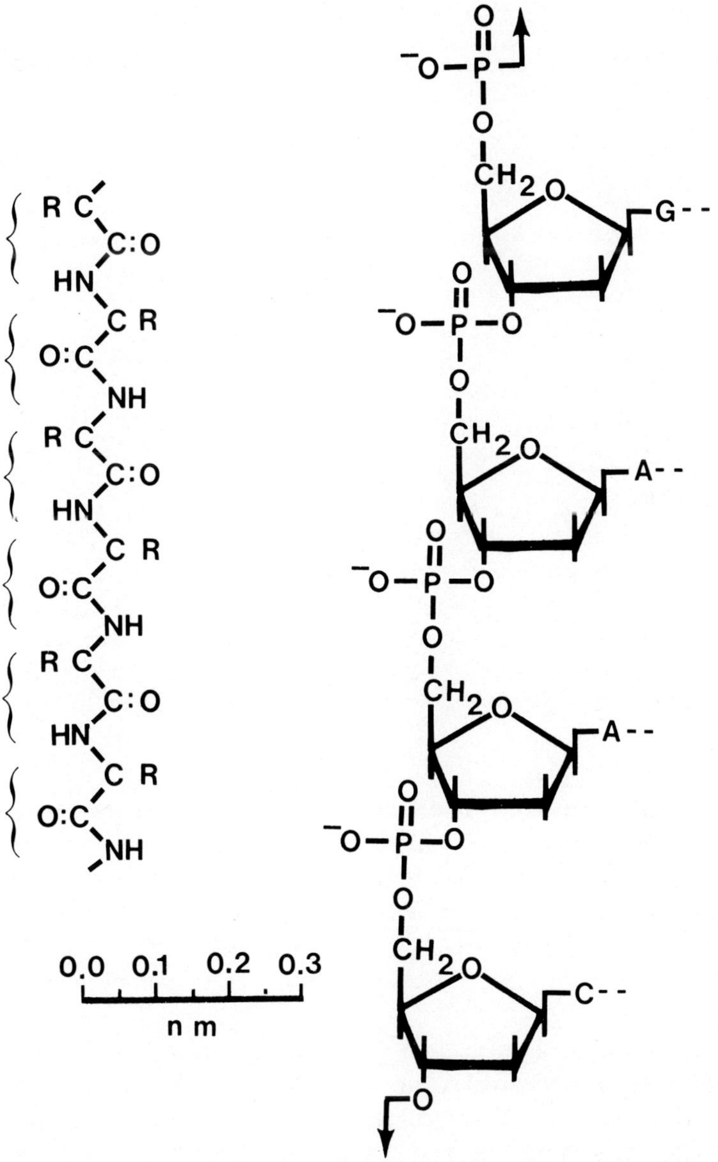

Figure 24. Interatomic distances in polypeptides and nucleic acids compared. A polypeptide chain is shown on the left. Brackets indicate individual amino acid residues. On the right is a polynucleotide chain. Note that the distance occupied by one nucleotide corresponds to the linear dimension of about three amino acids.

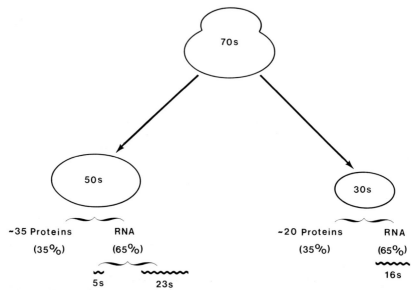

Figure 25. Structure of a bacterial ribosome. Two basic subunits make up the complete ribosome, having sedimentation coefficients of 50 and 30 svedberg units. These in turn are composed of protein and ribonucleic acid components as shown. (Note: The svedberg unit reflects the velocity a particle sediments in the ultracentrifuge, and is roughly proportional to molecular weight.)

percent G+C, in spite of the fact that the base composition of the DNA of the sources varied from 30 to 70 percent G+C. If the ribosomal RNA was a direct copy of the cell's DNA, then should not the rRNA reflect the base composition of the DNA, or at least the variations from species to species? The rRNA hypothesis was discarded and another messenger molecule was searched for.

Actually such a molecule was found in 1956, but was not recognized as such until some years later. Volkin and Astrachan discovered a new type of RNA in bacteriophage-infected bacteria. This RNA experienced a rapid turnover during the course of the infection (Fig. 26), and it was shown to have a base composition similar to that of the infecting phage DNA, but not that of the host bacteria. A few years later it was demonstrated that this RNA would hybridize with the DNA of the infecting

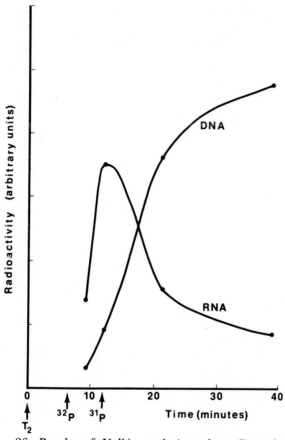

Figure 26. Results of Volkin and Astrachan. Bacteria were infected with bacteriophage T₂ at time zero. At about minute 6, radioactive ^{32}P phosphate was added to the cell suspension, and at minute 12, the cells were transferred to a medium containing nonradioactive ^{31}P. DNA and RNA were extracted from the infected cells during the course of the infection and measured for radioactivity. DNA radioactivity continues to rise even after the ^{31}P "chaser," indicating that ^{32}P-containing DNA precursors formed during the cells' exposure to the radioactive isotope are still being incorporated into polydeoxyribonucleotides, and are remaining in such. In contrast, the RNA in the cells rapidly loses its radioactivity as soon as the ^{32}P is removed from the medium, indicating that RNA made during the ^{32}P pulse is being recycled. (From L. Astrachan and E. Volkin, *Biochim Biophys Acta*, *29:*536, 1958. Reprinted with permission of Pergamon Press Ltd.)

phage, but not with the host DNA or with other phage DNA. This must have been the messenger molecule, and therefore it was named messenger RNA or mRNA. By 1960 it was found that mRNA is in all cells at all times, not just in phage-infected bacteria.

That DNA could in fact act as a template for the formation of mRNA was demonstrated by Ochoa, who received the Nobel prize in 1959 for carrying out such DNA-dependent RNA synthesis *in vitro*. It would appear that we now have the manner by which genetic information is transmitted from its record on the DNA to the site of protein synthesis, the ribosomes. The formation of mRNA from a DNA template is known as *transcription*.

But we still have not discussed how the language of nucleic acids is translated into the the language of proteins. Translation is a good term, for the problem is truly one of translation in the linguistic sense. DNA and RNA both consist of polymers of four nucleotides, whereas proteins consist of polymers of twenty amino acids. Thus, the information written in the nucleic acid base sequences is in a language of four letters, whereas that of proteins is in a language of twenty letters of an entirely different alphabet.

A one-to-one relationship between nucleotides and protein amino acids would of course not suffice, nor would a two to one ratio. The latter would only accomodate 4^2 or sixteen amino acids. A three-to-one ratio (that is, three nucleotides denoting each amino acid) would yield 4^3 or sixty-four possible combinations, more than enough for the twenty amino acids.

Recent studies of transcription have revealed that only one of the two strands of a DNA molecule is ever copied. In some cases, as in the bacteriophages T_4 and lambda, transcription may alternate from one DNA strand to another, whereas in other cases only one of the strands is copied. Thus, the overall base composition of an organism's complement of mRNA's could not mimic the composition of its DNA. The discovery of mRNA was aided by the unstable nature of it in phage-infected bacteria. This is not a common characteristic of mRNA, however, and many kinds of mRNA are highly stable. It is

therefore ironic that the arguments based upon these pur-
ported attributes of mRNA, its similar base composition to the
DNA of the cell, and its instability, that marked the pathway to
its discovery, were in fact specious.

Let us assume for the moment that genetic information is in
fact encoded in sequences of DNA purine and pyrimidine bases
taken three at a time. The genetic code or language thus
consists of three letter words in an alphabet of only four letters.
All possible combinations of the four letters taken three at a
time give us a tentative vocabulary of sixty-four words, which
when translated represent the twenty words (amino acids) of
the language of proteins. Of the experimental evidence offered
to support such a "triplet" code, the 1961 work of Crick and
coworkers is most often cited. Crick concentrated his attention
on one gene on the bacteriophage T4 chromosome known as
the B cistron of the rII region. This genetic region controls the
ability of the phage to lyse certain strains of *E. coli.* Crick took
advantage of a property of a class of compounds known as the
acridine dyes, examples of which are acridine yellow, acridine
orange, and proflavine. One of the characteristics of these
compounds is their ability to induce the deletion or addition of
single bases in the DNA chain. The mechanism of this action
will be covered in Chapter Four. If a bacteriophage suffers a
deletion or addition of a base in the rII region, it loses its ability
to form plaques on strain K of *E. coli.*

Suppose we simulate the message encoded in the normal
gene as "the sad man saw the fat cat." The action of an acridine
dye exposure causes the deletion of the *a* in sad so the message
would now read "the sdm ans awt hef atc at." (Remember, we
assumed the "reading frame" to be three bases. If one base is
removed, the reading frame is shifted over to fill the space left
by the missing base, resulting in this nonsense message.) A
second acridine treatment might result in the deletion of the *d*
in sdm to result in "the sma nsa wth efa tca t." Finally, a third
application of the dye might induce the deletion of the *n* in nsa
to produce "the sma saw the fat cat." We see we have regained
at least a partial sensible message after three base deletions.
This could only happen if the reading frame, that is, the length

of the genetic words, were three bases long, or multiples of three. The same outcome would result if we had dealt with three additions rather than three deletions; the reading frame would be restored as the reading mechanism passed the third added base. In both cases, with deletions or additions, the events had to occur in locations relative to the total genetic message such that enough sensible information was retained for the wild phenotype to be restored.

In Crick's experiment, the return to a semisensible message was reflected in the reacquisition of the ability of the bacteriophage to form plaques on strain K of *E. coli*. Crick published the results of his work in December of 1961, indicating that the genetic code appeared in fact to be a triplet code, and also suggesting that the code had to be degenerate. *Degeneracy* means that each of the twenty amino acids is represented by more than one nucleotide triplet. This would then account for the fact that there are sixty-four possible triplet codons, but twenty amino acids.

At about the same time Crick was establishing his hypothesis for a triplet genetic code, a team of American biochemists was making the first inroads into deciphering the code.

Establishment of the Code

In 1960 and 1961, Nirenberg and Matthaei developed a stable, cell-free, *in vitro* system of protein synthesis consisting of bacterial ribosomes, amino acids, ATP, RNA, and other enzymes and cofactors. In one experiment they added an amount of a simple, synthetic polynucleotide, polyuridylic acid, to the reaction mixture and found it stimulated the formation of a polypeptide consisting of just phenylalanine. The reaction was highly specific in that no other amino acid responded appreciably to the "poly-U," nor did other polynucleotides result in significant phenylalanine incorporation into polypetide. It was concluded that the poly-U acted as a messenger RNA for the formation of the polyphenylalanine, and therefore the genetic codon for phenylalanine was three uridylic acids (UUU).

The experiment was repeated with polyadenylic acid and polycytidylic acid, with the resulting incorporation into polymers of lysine and proline, respectively. Assuming a triplet code, we now have three codons translated, UUU for phenylalanine, AAA for lysine, and CCC for proline.

More complex mRNA's were produced in the laboratory, consisting of two, then three bases of known compositions. For example, a synthetic polyribonucleotide was produced with U, A, and C in a molar ratio of 1:1:4. That is, for every molecule of uracil in the chain, there was one molecule of adenine and four of cytosine. Assuming the bases were located on the chain in a random fashion, one can calculate the probabilities of the occurrences of all possible triplets. The probability of finding the triplet UUU would be the result of multiplying the probablity of finding one U (1/6) times the probability of finding a second (1/6), times the probability of finding a third U (1/6), or (1/6)(1/6)(1/6)=(1/216). Finding two C's and an A would be one chance in (4/6)(1/6)(1/6) or 4/216. Table VII lists the nine possible combinations of the three bases in our example and their probabilities of occurrence. Note that at this stage of the work, the *order* of the bases as they might appear in the RNA chain is not known, only that we have two C's and an A, or two A's and a U, etcetera.

If this random RNA chain is then used as a mRNA in a cell-free protein synthesizing system as described above, and

TABLE VII
PROBABILITIES OF OCCURANCE OF TRIPLETS OF U, A, AND C IN
SYNTHETIC RNA WHEN THEIR MOLAR RATIOS ARE SET AT 1:1:4

Triplet*	Probability	Normalized probability
C_3	.296	100
C_2A	.074	25
C_2U	.074	25
A_2C	.018	6
U_2C	.018	6
UAC	.018	6
A_2U	.0046	0.5
U_2A	.0046	0.5
U_3	.0046	0.5
A_3	.0046	0.5

*C_3 denotes a triplet of three C's. C_2A denotes 2 C's and an A, the order of which may be CAC, CCA, or ACC, etc.

the amino acid composition of the resulting polypeptide is determined, one begins to fit pieces of the genetic code puzzle together. Much of this work was carried out by Severo Ochoa and coworkers from about 1961 to 1963. Table VIII lists the composition of the polypeptide that was produced from the mRNA described in Table VII. Proline appears to be the most frequent amino acid, followed by histidine and threonine. A third level of frequency is occupied by asparagine, glutamine, phenylalanine and tyrosine. The frequency levels exhibited by these seven amino acids approximate the three upper frequency levels exhibited by the mRNA triplets. We can thus assign tentative triplets to the amino acids within each frequency level. By repeating the experiment with mRNA's of different composition, it eventually becomes possible to assign more specific triplets to each amino acid.

TABLE VIII

COMPOSITION OF THE POLYPEPTIDE SYNTHESIZED *IN VITRO* WITH THE mRNA DESCRIBED IN TABLE VII*

Amino acid	Moles (Normalized)	Possible triplet codons
Proline	100	C_3 (CCC)
Histidine	36 }	C_2U (CCU, CUC, UCC)
Threonine	38 }	C_2A (CCA, CAC, ACC)
Asparagine	10	
Glutamine	9	
Phenylalanine	6 }	A_2C (AAC, ACA, CAA)
Tyrosine	5 }	U_2C (UUC, UCU, CUU)
Isoleucine	7	UAC (UAC, ACU, CAU, AUC)

*Data from Speyer, J. F., Lengyel, P., Basilio, C., Wahba, A. J., Gardner, R. S., and Ochoa, S.: *Cold Spring Harbor Symp Quant Bio, 28:*559, 1963.

A consequent approach was to synthesize mRNA's with nonrandom, that is, known sequences, such as ACACACACAC. A mRNA with this sequence offers but two possible triplets, ACA and CAC. Thus some of the amino acids whose assignments were known only as A_2C and C_2A from the random nucleotide work now could be given definite assignments. Additional evidence for codon assignments came from yet another route, that of determining what isolated trinucleotides would bind what specific amino acids to ribosomes. This technique was devised by Nirenberg and Leder (1964) and made it possible to deduce many additional codons. Much of the excitement in seeing the various pieces of the puzzle fall into place from so many laboratories was expressed in Crick's introduction to the 1966 Genetic Code volume of the Cold Spring Harbor Symposia. Crick traveled the country collating the results of various workers, and eventually drew up the codon listing shown in Figure 27. The structures of the twenty common amino acids are shown in Table IX.

Examination of the codon assignments immediately confirms Crick's early suggestion of a degenerate code. Most of the amino acids possess more than one codon, the only exceptions being methionine and tryptophan. A binary periodicity is evident in which each amino acid except methionine and tryptophan appears with a cluster of either two, four, or six codons. This cluster effect is brought about by the fact that the first two triplet bases for each amino acid are almost always the same. The phenylalanine codons begin UU-, isoleucine, AU-, serine, UC-, and so forth. The significance of this is that the third member of a triplet codon may possess a lower level of importance for distinguishing its triplet from other triplets. For example, the codon for tyrosine may be considered as UApyrimidine, for glycine, CApurine. For methionine and tryptophan, however, the distinction between G and A does seem to make a difference. We will continue this line of thought in a later section. A final observation for the time being involves the triplets UAA, UAG, and UGA. These do not code for any amino acid, and in reality act as a kind of punctuation or termination signal for the sequence of triplets that codes for a given polypeptide.

We have now traveled to the point where we have in mind the manner by which genetic information for the structure of a particular protein is encoded in the base sequence of the DNA making up a specific gene, and how that sequence is copied by forming a complementary molecule of mRNA. The mRNA now has spelled out in its structure the sequence of amino acids that make up the protein.

2nd →	U	C	A	G	3rd ↓
1st ↓					
U	PHE	SER	TYR	CYS	U
					C
			1	3	A
	LEU		2	TRP	G
C	LEU	PRO	HIS	ARG	U
					C
			GLN		A
					G
A	ILEU	THR	ASN	SER	U
					C
			LYS	ARG	A
	MET				G
G	VAL	ALA	ASP	GLY	U
					C
			GLU		A
					G

Figure 27. The genetic code. Listed are all the sixty-four possible triplet codons along with their amino acid assignments. Three codons, UAA, UAG, and UGA (shown as 1, 2, and 3) are nonsense codons and are used as punctuation or termination signals in protein synthesis. Key to the abbreviations will be found in Table IX.

TABLE IX 67

THE 20 COMMON AMINO ACIDS

General structure:
$$R-\underset{\underset{O}{\overset{\|}{C}}}{\overset{\overset{NH}{|}}{C}}HCOH$$

Amino acid	*Abbreviation*	*R–*
Glycine	gln	$H-$
Alanine	ala	CH_3-
Serine	ser	CH_2OH-
Threonine	thr	CH_3CHOH-
Valine	val	$(CH_3)_2CH-$
Leucine	leu	$(CH_3)_2CHCH_2-$
Isoleucine	ile	$CH_3CH_2CH(CH_3)-$
Cysteine	cys	$HSCH_2-$
Methionine	met	$CH_3SCH_2CH_2-$
Phenylalanine	phe	$-CH_2-$
Tyrosine	tyr	HO $-CH_2-$
Histidine	his	$-CH_2-$
Tryptophan	trp	$-CH_2-$
Proline	pro	$CH_2CH_2CH_2NHCHCOOH$
Aspartic acid	asp	$HOOCCH_2-$
Asparagine	asn	H_2NCOCH_2-
Glutamic acid	glu	$HOOCCH_2CH_2-$
Glutamine	gln	$H_2NCOCH_2CH_2-$
Arginine	arg	$HN:CNH_2NH(CH_2)_3-$
Lysine	lys	$H_2N(CH_2)_4-$

POLYPEPTIDE CHAIN FORMATION

The Adaptor Molecule

We now turn to the mechanisms that make it possible for a cell to convert the genetic information transcribed onto the mRNA into a protein molecule. The early notion that the mRNA may act as a direct template for protein formation was quickly dispelled on many of the same grounds as was the DNA template hypothesis. About 1960, Crick formulated the hypothesis that the relationship between amino acids and mRNA was mediated through what were referred to as "adaptors." It was these small molecules that allowed specific amino acids to marshal along the mRNA in the sequence commanded by the sequence of nucleotides, to be polymerized eventually into a polypeptide chain.

It had been observed that about 10 to 20 percent of the RNA of a cell was made up of a low molecular weight (25,000 to 30,000 daltons) fraction which was designated as "soluble RNA" (sRNA). The base composition of sRNA was uniformly around 58 percent G+C regardless of its source, but one of its most remarkable attributes was that amino acids were frequently associated with it. That is to say, routine attempts to purify sRNA failed to remove traces of amino acids from it. Molecules of sRNA that were free of amino acids could be linked with, or "charged" with amino acids in a system that involved enzymes implicated in protein synthesis. It soon became clear that the sRNA molecules were the adaptors that Crick had postulated. Since these molecules have the responsibility of transferring amino acids from their free state in the cytoplasm to the active environment of the ribosomes, they were renamed transfer RNA, or tRNA.

Because of their relatively short chain length (less than 85 nucleotides compared with thousands for mRNA and millions for chromosomal DNA) it has been possible to determine the primary nucleotide sequences of at least sixty different tRNA's, every amino acid having one or more tRNA's specific for it.

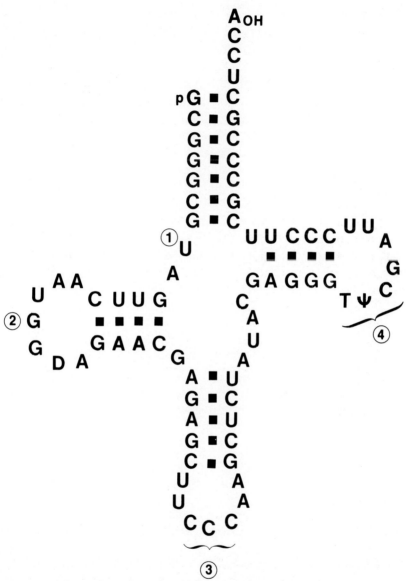

Figure 28. Structure of one of the glycine tRNA's from *E. coli.* Footnotes: 1. The uracil at this location appears to be 4-thiouracil. 2. In some molecules from this organism, this guanine occurs as 2'-O-methyl guanine. 3. This triplet (CCC) is the anticodon for this tRNA and corresponds to the glycine codon GGG. 4. One of many constant regions among tRNA's, this triplet appears to be the attachment site to the ribosome. Other symbols: D = 5, 6-dihydrouracil; ψ = pseudouracil (See Table X). (From Hill, C. W.: *J Biol Chem*, 248:4252, 1973)

The primary structures of all of the known tRNA's can be arranged in a cloverleaf configuration such as the tRNA for glycine depicted in Figure 28. About half of the nucleotides can participate in Watson-Crick base pairing to produce partial double-strandedness. Every tRNA terminates with -CCA, with the 3' hydroxy group of the adenosine ribose being the attachment site for the specific amino acid. Careful examination of the bases of the tRNA molecules reveals a startling assemblage of unusual purines and pyrimidines. Over forty such bases have been identified in tRNA from various sources, some of which are listed in Table X. Between 10 and 20 percent of the bases making up tRNA molecules are of these minor types. The most extraordinary of the list must be the sulfur-containing pyrimidines such as thiouracil and 2-thiocytidine. Thymine, normally thought of as only being found in DNA, is also on the list. The role of these unusual bases is unknown.

TABLE X
SOME UNUSUAL PURINE AND PYRIMIDINE BASES FROM tRNA

Common base	Found in transfer RNA
Guanine	1-methyl-
	N(2) dimethyl-
	2'-0-methyl* guanylic acid
Adenine	N(6)dimethyl-
	N(6)isopentyl-
Uracil	5-methyl- (=thymine)
	4-thio-
	pseudouridylic acid[†]
Cytosine	5-methyl-
	N(6)acetyl-
	2-thio-
	2'-0-methyl* cytidylic acid

*Methyl group on the number 2 oxygen of the ribose
[†]Ribose attached to the number 5 position of uracil instead of number 3
See Figure 3 for the structures of the common bases.

Newly formed tRNA does not contain these minor bases, nor does it possess the -CCA at the 3' end. Following their formation from specific tRNA genes on the DNA, the tRNA molecules are modified by numerous enzymes to result in the structures that participate in protein synthesis.

At the end of the middle loop of a tRNA molecule three

nucleotides form the *anticodon*. It is this triplet that permits the tRNA molecule, charged with its specific amino acid, to match up on the mRNA with the complementary codon that designates that amino acid. As each succeeding amino acid is lined up along the mRNA, it is enzymatically joined to the preceeding amino acid by a peptide bond, and ultimately a polypeptide chain is formed. This gross oversimplification of the reaction hides the fact that over 100 reactants and cofactors are involved in it. A somewhat more detailed discussion now follows.

Amino Acid-tRNA Complex

The enzymes that aid in the formation of the key precursors of polypeptide synthesis, the amino acid-tRNA complexes, are known as amino acid-tRNA synthetases. There are at least twenty different synthetases, one for each of the twenty naturally occurring amino acids. For example, alanine-tRNA synthetase first catalyzes the formation of an alanine adenylic acid complex from free alanine and ATP. The subsequent step, also catalyzed by the same enzyme, is the exchange of the adenylate from the ATP for the terminal adenylate of the alanine-tRNA, forming an alanine-alanine tRNA complex. These steps are shown in Figure 29.

Initiation of Protein Synthesis

Protein synthesis occurs at the ribosomes, where initiation of synthesis involves the formation of a complex between a mRNA molecule, a tRNA molecule that is charged with formyl-methionine, and a 30s ribosome subunit (Fig. 30). At this point, a 50s ribosome subunit links to the complex to complete the ribosome. The ribosome now passes along the mRNA chain in a rachet-like manner, stopping at each triplet codon to enable a specific tRNA molecule, charged with its amino acid, to align itself anticodon to codon, along the mRNA. Molecules of

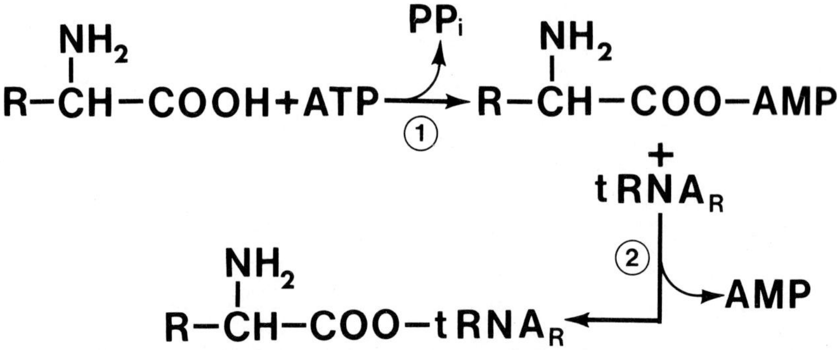

$$\underset{\underset{\text{R--CH--COOH+ATP}}{|}}{\overset{\text{NH}_2}{}} \xrightarrow[\textcircled{1}]{\overset{\text{PP}_i}{}} \underset{\underset{\text{R--CH--COO--AMP}}{|}}{\overset{\text{NH}_2}{}}$$

$$+$$

$$\text{t RNA}_\text{R}$$

$$\underset{\underset{\text{R--CH--COO--t RNA}_\text{R}}{|}}{\overset{\text{NH}_2}{}} \xleftarrow[\textcircled{2}]{} \text{AMP}$$

Figure 29. Steps to the formation of amino acid-tRNA complex. In step 1, an amino acid is activated by reaction with ATP to form an amino acid-AMP complex. The latter then combines with a tRNA specific for the amino acid to form the amino acid-tRNA complex. Both steps 1 and 2 are catalyzed by the same enzyme, an amino acyl tRNA synthetase specific for the amino acid involved.

guanosinetriphosphate (GTP) appear to supply the energy for this movement. Each amino acid is then covalently attached to the amino acid that came before it through a polypeptide bond and is simultaneously released from its tRNA. The sequence of codons on the mRNA that designates a particular polypeptide ends with one or more of the nonsense triplets UAA, UAG, or UGA. The presence of these triplets signals termination factors that release the nascent protein from the mRNA ribosome complex. If this also happens to be the end of the mRNA molecule, it too is released from the complex, all of the subunits of which part to regroup and initiate the cycle again. In some instances, however, a mRNA molecule may carry information for the formation of two or more distinct proteins.

One of the most striking examples of such a *polygenic* mRNA is in the formation of the enzymes involved in histidine biosynthesis in *Salmonella*. A single mRNA molecule has encoded the amino acid sequences of nine enzymes, the nucleotide series for each protein presumably being separated from one another by termination and initiation codons.

Nearly all of what we know of the details of protein synthesis has been derived from studies of bacteria, either normal or

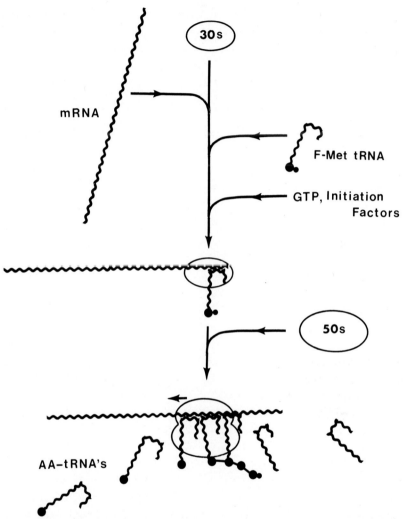

Figure 30. Steps in protein synthesis. A mRNA molecule and the 30s component of a ribosome combine with formyl methionine tRNA and other cofactors, followed by a 50s ribosomal subunit to complete the ribosome. Guided by specific codon-anticodon interactions, amino acids are then sequenced according to the genetic message encoded on the mRNA. Polypeptide bonds are formed enzymatically, and the resulting protein is completed.

bacteriophage infected. It appears that all bacterial protein synthesis is initiated with an N-formyl methionine. About 95 percent of all bacterial proteins begin with either a methionine, alanine, serine, or threonine, with methionine representing some 40 percent of the total. Thus, following the formation of a protein chain, the cell has the capability of removing either the formyl group, the methionine, or both. Enzymes to carry out such reactions have been detected in *E. coli* and *Bacillus subtilis.*

CONTROL OF PROTEIN SYNTHESIS

Inducible Enzymes

On two occasions we have repeated the supposition that bacteria as a group have the capability of producing 4,000 or more different proteins. For a given species, such as *E. coli* or *Thiobacillus thiooxidans,* the figure may be no more than a thousand or so. Although a bacterium certainly has the genetic capacity to produce 1000 to 2000 different proteins, it is not at all clear that it would have the energy capability to produce them all simultaneously in significant quantities. In fact, for some time we have known that this does not happen. In the early 1930's the Finnish bacteriologist Karström found that a specific species of bacterium varied in its capability to produce certain enzymes depending on the medium in which it was grown. If a given substrate were present in the medium, then the bacteria would demonstrate catabolic activity towards that substrate. If the substrate were absent, little or no specific enzyme activity could be found. Karström called these enzymes *adaptive.* It was later shown that the presence or absence of a given enzyme activity was due to the cells' ability to turn the formation of that enzyme on and off. We now refer to a specific enzyme as being *inducible* where the presence of the substrate (or an analog of it) induces the formation of that enzyme. This is in contrast to so-called *constitutive* enzymes which are always produced by the cells regardless of the absence or presence of inducers.

The kinetics of inducible enzyme formation have been intensely studied by many workers, including the French microbiologists and Nobel laureates Jacob and Monod, who showed (Fig. 31) that on induction an inducible enzyme may

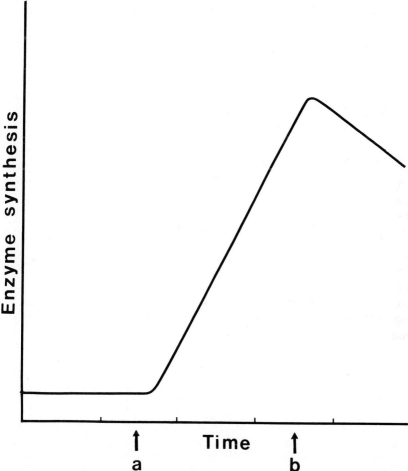

Figure 31. Kinetics of inducible enzyme formation. The activity of a given inducible enzyme is measured in a bacterial culture that is initially grown in the absence of the enzyme's inducer. At time (a.) inducer is added to the medium, and at time (b.), the cells are removed from the medium and resuspended in a medium lacking the inducer. At this point, synthesis ceases and enzyme levels eventually drop to that observed before inducer was added.

quickly represent over 6 percent of the cells' total protein synthesis. These workers also established the mechanism by which bacterial cells control the formation of inducible enzymes.

Two or three inducible enzymes are involved in the initial catabolism of the disaccharide lactose in *E. coli.* The first enzyme is β-galactoside permease, a protein that aids in the passage of galactosides such as lactose into the cell. The second is β-galactosidase, which splits lactose into glucose and galactose. The final enzyme is β-galactoside transacetylase. The genes responsible for the synthesis of these enzymes are clustered together in a specific region of the bacterial chromosome and are known as the *y, z,* and *a* genes, respectively (Fig. 32). A number of mutants of *E. coli* had been isolated by Jacob and Monod that had lost their ability to ferment lactose, as evidenced by their lactose-negative colonies on lactose-EMB agar. Through genetic means, which will be discussed in detail in later chapters, it was found that most of the mutations mapped in one or another of the structural genes *y, z,* or *a.* A mutation in the *z* gene, for example, caused a loss of the ability to produce β-galactosidase, although normal levels of the other two enzymes were formed. Another class of mutants demonstrated a remarkable change. In these strains, all three

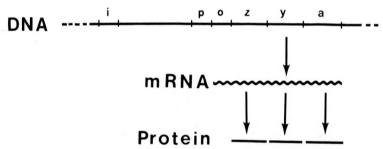

Figure 32. The lactose operon of *Escherichia coli.* Regulatory genes *p* and *o,* plus structural genes *z, y,* and *a,* make up the genetic region of coordinated transcription involved with the catabolism of lactose known as the lactose operon. The *i* gene, while not part of the operon, is responsible for the formation of repressor, which blocks transcription of the operon (Figure 33).

enzymes were no longer inducible, but constitutive. That is to say, the enzymes were produced in normal concentrations even in the absence of lactose. The mutations that these cells suffered mapped in a location outside of but near the structural genes z, y, and a. The site of these mutations was assumed to be a gene that controlled the inducibleness of the structural genes and was hence called the i gene.

The i gene falls into a second category of genes, *regulatory genes*. Up to this point we have only considered *structural genes*, those sequences of DNA that code for specific enzymes or structural proteins that generally carry out their roles remote from the chromosome.

The Operon

The regulatory genes, on the other hand, while they may also produce a polypeptide, act on or otherwise directly affect other genetic regions of the cell's chromosome. Thus, when the i gene experienced a mutation, it lost its control over all three of the lactose genes.

Other regulatory genes in the lactose region are the *promoter* (*p*) and the operator (*o*). The promoter region is the site of initial RNA polymerase binding and of initiation of transcription. The entire genetic locale, including the p, o, z, y, and a genes, but not the i gene, has been named the lactose *operon*. An operon is defined as a genetic region of coordinated transcription.

Repressors

What is the nature of the control of the i gene? In a series of experiments by Jacob, Monod and others, it was revealed that the level of control of the i gene was transcription. That is, the i gene, in its wild (unmutated) state and in the absence of inducer, prevented the transcription of mRNA from the z, y, and a genes. In the presence of lactose or other inducer molecule, however, the *i* gene released its repression of the structural genes, transcription proceeded normally, and ultimately the enzymes were formed. The *i* gene produces a protein *repressor,* it has subsequently been found, and it is the repressor that blocks transcription of the structural genes by binding with the *operator* region (Fig. 33). The operator, or "*o*" region, is situated immediately adjacent to the structural genes and is presumably the location through which RNA polymerase must pass in transcription. Alternatively, the presence of the repressor bound to the *o* region may simply prevent initial binding of the RNA polymerase to the adjacent *p* site.

On induction, that is, in the presence of lactose, the inducer reacts with the repressor (Fig. 34), transcription is released and enzymes are ultimately produced. When the level of inducer (substrate) drops below a certain level, repressor is again available to react with the *o* region and transcription is stopped. The lactose operon is thus referred to as one being under *negative control,* in which the product of a regulator gene turns the operon off.

The isolation of the actual repressor molecule was based on the assumption that it reacted directly with the inducer. In 1967 Gilbert and Müller-Hill succeeded in isolating lactose repressor by incubating *E. coli* cells with a radioactive nonmetabolizable inducer, isopropylthiogalactoside. On concentrating the cell extracts they found a tetrameric protein with a molecular weight of about 1 to 2 x 10^5 that was bound firmly to the inducer. The purified protein has subsequently been shown to

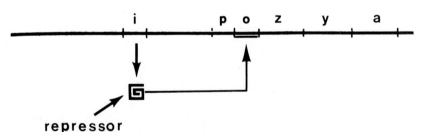

repressor

Figure 33. Repression of the lactose operon. The *i* gene elicits the formation of a repressor molecule that binds to the *o* gene and prevents transcription of the structural genes of the operon.

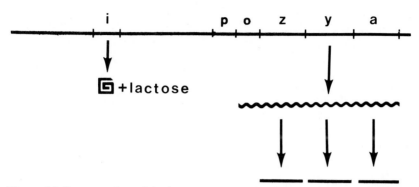

Figure 34. Derepression of the lactose operon. The inducer, lactose, combines with the repressor and prevents it from binding to the *o* gene. The *o* gene is now free to initiate transcription of the structural genes.

bind firmly to isolated lactose operon DNA and to prevent *in vitro* transcription of the *lac* operon. In the presence of suitable inducer, however, the *in vitro* transcription proceeds normally.

Allosterism

The behavior of the repressor protein is attributed to a property known as *allosterism.* An allosteric protein is one that possesses two reactive sites. If one of the sites reacts with an *effector,* the activity of the second site is altered. The second site, even though it may be remote from the first on the molecule, may either be activated or inhibited. In the case of the lactose repressor (Fig. 35), reacting with the inducer (lactose) at one site inhibits the second site from binding with the operator. How remote sites on a protein molecule may influence one-another has been explained in terms of reformations of the tertiary and quaternary structures of the molecule by the presence of the effector.

As implied in the previous paragraph, control systems exist in which effectors may activate an allosteric regulatory molecule, which in turn would initiate transcription. Figure 36 depicts such a system in which no transcription is being carried out, hence no enzymes are formed. The inactive *apo-activator*

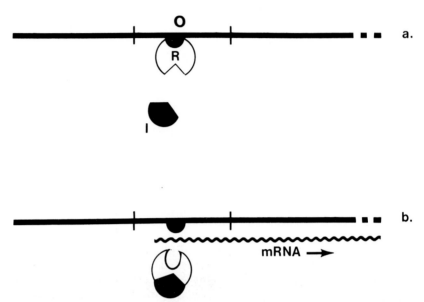

Figure 35. Hypothetical mechanisms of repression and derepression. When the operon is repressed (a.), the allosteric repressor molecule (R) is bound to the *o* gene at a specific binding site, blocking transcription. An inducer molecule (I) can combine with the repressor, altering its ability to bind to the *o* gene (b.), and allowing transcription to proceed.

reacts with a specific inducer, is activated, and triggers the transcription of the structural genes. In bacteria the best example of such a *positive control* system is that concerning arabinose metabolism, in which three structural genes are under the control of a regulatory gene. The apo-activator is activated by arabinose, the known inducer of the operon, which then reacts with the operon to initiate transcription.

Another type of coordinated transcription is one frequently referred to as *end product repression,* in which the presence of the effector brings about the repression, or turning off of the operon rather than its activation. In this case the effector is frequently the end product of the pathway controlled by the operon (Fig. 37) or a derivative of it. We thus have a type of feedback mechanism in which the excess accumulation of the end product brings about the repression of the operon

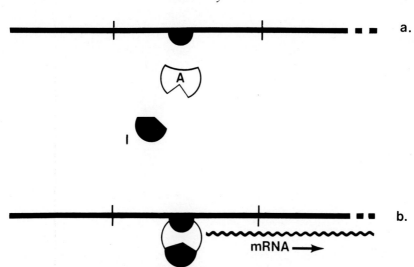

Figure 36. Induction of an operon under positive control. An operon (a.) is not carrying out transcription until an inducer (I) combines with an apo-activator (A), which in turn can now bind to the operon and initiate transcription (b.).

responsible for the formation of that product.* In the formation of the amino acid histidine in *Salmonella,* histidine (actually a derivative of histidine-tRNA) appears to activate an apo-repressor which ultimately brings about the cessation of histidine biosynthesis by repressing the histidine operon.

What happens when a metabolic path is branched, with several end products? Frequently, a mechanism known as *multivalent repression* comes into play here in which each of the end products must be at maximum concentrations before repression is activated.

Among microorganisms and higher plants and animals, there are over two dozen examples of operons such as we have just described, and no doubt many more are yet to be discovered. Thus, nature has evolved complex systems which

---

*This is not to be confused with classical *feedback inhibition* in which the end product of a metabolic pathway reacts directly with an enzyme involved in an early step in the pathway and inhibits its activity.

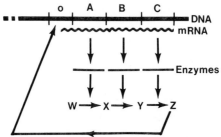

Figure 37. End product repression. An operon with structural genes A, B, and C is responsible for the formation of enzymes that catalyze the biosynthesis of Z from W. An accumulation of Z can bring about the repression of the operon by Z or a derivative of it combining with the operator gene, usually in combination with an apo-repressor.

afford organisms the ability to economize on energy and materials, a philosophy that the human organism might well adhere to more closely.

INDUSTRIAL APPLICATIONS

As you are aware, hundreds of compounds are produced commercially, sometimes in multiton quantities, by the activities of microorganisms. Familiar examples are citric acid, lysine, vitamin B-12, dextran, and of course the antibiotics. Knowledge of the workings of metabolic control systems has had a considerable impact on the economics of industrial fermentations, in which success is highly, if not solely, dependent upon yield. Maximum yield is based on the accumulation of the desired product in the medium, but we have just learned in previous sections that microorganisms have evolved mechanisms that prevent the wasteful overproduction of metabolites. Various means have thus been developed to by-pass these control systems, at both the transcription level and the enzyme level, in order to achieve commercially acceptable yields.

An example of disconnecting a repression step is found in the case of isoleucine production in *E. coli* in which pantothenic acid, leucine, valine, and isoleucine cooperate in the multivalent repression of three enzymes involved in their biosynthesis (Fig. 38). However, valine also acts outside of this multivalent repression system, in that an excess of this amino acid alone inhibits growth. In isolating mutants that were resistant to inhibition by valine, it was found that these strains excreted large amounts of isoleucine. Genetic analysis shows that these strains appear to have suffered a mutation in a regulatory gene in the isoleucine operon which afforded them resistance towards repression by the four metabolites involved. The isolation of valine resistant cells offered a fortuitous means of selecting for such repressionless mutants.

To attack control systems at the transcription level is exceedingly difficult, mainly because of lack of knowledge of specific operons. A more common approach is to upset a feedback inhibition step, such as in the production of ornithine by *Corynebacterium glutamicum* (Fig. 39). Glutamate is eventually converted to arginine, with ornithine as one of the intermediates. The end product, arginine, is the controlling metabolite, influencing the second step in the pathway through feedback inhibition. Mutants have been isolated which have suffered a block in step 6 and are unable to convert ornithine to citrulline. By supplying the organisms limiting quantities of arginine, the organisms continue to grow and make ornithine. This intermediate, not being converted to the next metabolite in the pathway, accumulates in large amounts in the medium.

84

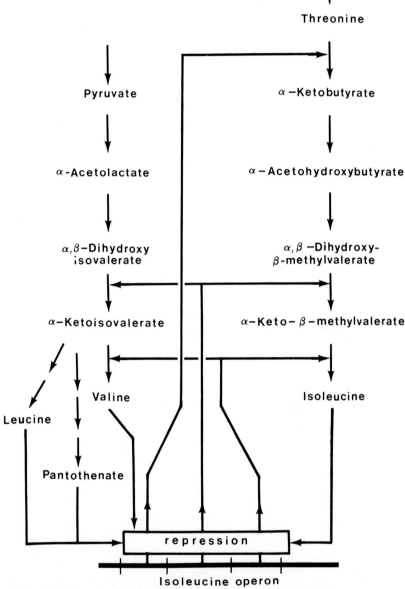

Figure 38. Example of a disconnected repression step. The isoleucine operon (bottom of figure) controls the formation of three enzymes involved in the formation of leucine, pantothenate, valine, and isoleucine, all of which cooperate in the multivalent repression of the operon. Accumulations of valine alone also efficiently repress the operon, but mutants have been isolated that are not inhibited by high concentrations of valine. These mutants have apparently experienced a loss of the repression mechanism, and as a result, accumulate large quantities of isoleucine in the medium. (From Adelberg, E. A.: *J Bacteriol, 87:*566, 1964).

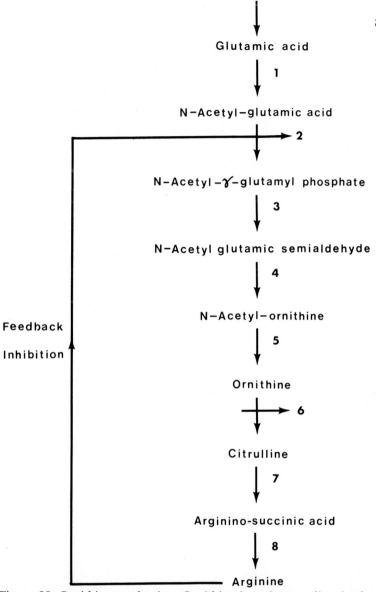

Figure 39. Ornithine production. Ornithine is an intermediate in the formation of arginine from glutamate. Through feedback inhibition the organism is prevented from accumulating large amounts of arginine or any of its precursors. A mutant has been isolated, however, that has suffered a block in the conversion of ornithine to citrulline. This mutant, if supplied amounts of arginine just sufficient for growth, will accumulate large quantities of ornithine in the medium, for the concentration of arginine supplied is not high enough to bring about feedback inhibition of this pathway. (From Udaka, S., and Kinoshita, S.: *J Gen Appl Microbiol, 4:*283, 1958).

Chapter Four

MUTATIONS I

MUTATIONS IN BACTERIA

GREGOR MENDEL BROUGHT to our attention the existence of *alleles*, or alternate forms of a given gene. The gene that controlled stem length in Mendel's garden peas, for example, exhibited two forms: one that produced *short* stems and one that produced *long* stems. How did these alternate forms arise? Biologists frequently observed changes appearing spontaneously in populations of plants and animals. Another nineteenth century figure, Hugo de Vries, made many such observations among his object of study, the primrose. A few percent of the seedlings of every crop of primroses exhibited forms differing from the parental type, such as plant size, or leaf shape. De Vries was struck by the constancy of appearance of the aberrant forms, and named the sudden changes "mutations." As you recall from Chapter One, we now define mutations as changes in the molecular structure of a gene. While many of the mutations observed and so-called by de Vries are now known not to fit this strict definition, he is still given the credit for establishing the term. You may also recall from Chapter One that this Dutch botanist was one of the three men who, in 1900, rediscovered the work of Mendel.

Sudden changes were routinely observed in all organisms, such as mice, rabbits, fruit flies, and bacteria. The technique of pure culture developed by Koch in 1881 enabled bacteriologists to observe the characteristics of a single species of bacterium over many generations, and as a result, to discover spontaneous changes in members of the culture. For example, in any culture of the pigmented bacterium *Serratia* growing on agar medium, there were always a few colonies appearing that were colorless (Fig. 40). Subculturing the colorless colonies demonstrated that the bacteria had permanently lost the ability to form the pigment, all other characteristics remaining normal. The Dutch bacteriologist Beijerinck, on reading of the studies of mutations of his countryman de Vries, proposed in 1900 and for some years thereafter that the heritable changes observed in bacteria were also due to mutations.

But Beijerinck's proposal was nearly fifty years ahead of its time. Biologists were willing to accept the concept of gene mutations in higher plants and animals, but not in bacteria. At least two reasons were expressed for this position. First, it had been established that chromosomes were the carriers of genetic information, and yet no such structures could be demonstrated unequivocably in bacteria at this time. Secondly, an entire bacterial culture frequently acquired the new characteristic overnight, a pace many believed was much too rapid for a genetic process. This latter phenomenon was particularly evident in the case of bacteriophage-resistant variants. Very soon after the discovery of bacteriophages in 1915, it was observed that if a broth culture of susceptible bacteria were inoculated with a small number of phage particles and incubated, by the following day the culture would be turbid with bacteriophage-resistant bacteria.

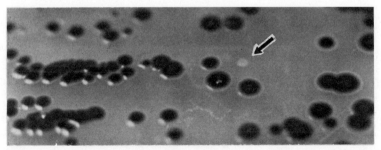

Figure 40. Pigmentless mutants. Occasional colorless colonies of pigmented bacteria such as *Serratia marcescens* will appear on agar plates. The pigment loss is due to a mutation that blocks the formation of the pigment.

What was overlooked was the recognition that a bacterial culture is made up of millions of individual cells, each undergoing a division perhaps every thirty minutes or so. Thus, a broth culture of bacteria may go through many generations in a twenty-four-hour period offering numerous opportunities for mutations to occur. With the selective pressures one can impose upon a bacterial culture, such as lethal bacteriophage, rapid changes in the culture are now easily understood. This understanding was not widespread in the early decades of this century, however. Most bacteriologists believed bacteria experienced changes as result of adaptation to environmental pressures, a concept generally attributed to the French biologist Lamarck (1744-1829). The Lamarckian doctrine was fairly well disposed of as it applied to higher plants and animals by the turn of the century, but as Werner Braun has written, bacteriology had the dubious distinction of being known as the last stronghold of Lamarckism. Remnants of it appeared in the bacteriological literature as late as the 1950's. However, in the early decades of this century a few did hold to the theory that bacteria had genes and were capable of experiencing spontaneous mutations and probably other genetic processes as well. Among these persons were such prominent microbiologists as Neisser (1906), Massani (1907) and Burnet (1929).

Luria-Delbrück Fluctuation Test

The first significant evidence for the presence of spontaneous mutations in bacteria was presented by Luria and Delbrück in 1943. Delbrück had been trained as a physicist, but in the 1930's he was attracted to biology, more specifically to the application of physical principles to gene structure and cell reproduction. No known system appeared simpler or easier to handle in the laboratory than bacteriophage reproduction, and it was here that Delbrück pioneered, later to be joined by Luria, a physician specializing in medical physics and radiology. Their experiments involved the use of bacterial hosts that had acquired resistance to infection by certain phage types. The origin of the resistant bacteria intrigued Luria for a number of years until, in 1943, he devised an experiment that would differentiate resistance acquired by adaptation versus resistance acquired by spontaneous mutation.

The experiments consisted of the preparation of a series of small broth cultures of a phage-susceptible strain of *E. coli* (Fig. 41). Following overnight incubation, an accurately measured aliquot of each culture was spread onto the surface of a nutrient agar plate that had previously been spread with a suspension of coliphage. The plates were incubated, and on the following day examined for colony formation. Because of the presence of the phage, most of the bacteria spread onto the plates had been lysed. However, several colonies did emerge on the plates that appeared to be resistant to the phage. (The colonies may be subcultured to confirm their members to be resistant and not contaminants). The pivotal question is whether these colonies arose as a result of spontaneous mutations that occurred in a few bacteria during the overnight incubation period that afforded the cells resistance to phage infection, or they arose as a result of an adaptive process in which a constant fraction of any susceptible population of bacteria will always prove to be resistant. The exact mechanism of the latter process, whether through classic Lamarckian adaptation or some other mechanism, is immaterial at this point. That is to say, we are only concerned with the question of whether spontaneous mutations do or do not occur in bacteria.

By definition, spontaneous mutations would occur randomly in a population. One cannot predict when a given cell will experience a mutation. Thus, suppose in a rack of culture tubes in the Luria-Delbrück experiment, five minutes after the rack is placed in the incubator, a cell in tube number 1 experiences a mutation that affords it phage resistance. At the end of x generations there will be 2^x offspring of the mutant cell, all of which will be phage resistant. Of course there would be more resistant cells if additional mutations occurred during the incubation period. But suppose in tube number 2 a mutation to phage resistance did not occur until the following day, say five minutes before the rack was removed from the incubator. Tube number 2 now contains one resistant cell. Due to the

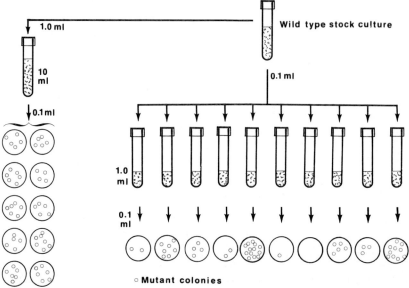

Figure 41. The Luria-Delbrück experiment. Ten culture tubes, each containing 1 ml of broth, were inoculated with equal numbers of wild type bacteria. Following a 24-hour incubation, each tube was assayed for the number of phage resistant variants by plating an aliquot onto an agar plate seeded with bacteriophage. Luria and Delbrück reasoned that if the resistant variants were the result of adaptation, equal numbers of colonies would have formed on each plate. If spontaneous mutations were the cause of the few resistant bacteria, significant fluctuations in the numbers of colonies appearing on the plates would be observed. Which hypothesis is supported? The tube containing 10 ml of broth (left side of figure) was assayed in 10 replicate samplings to determine the degree of fluctuations due to random error alone.

randomness of mutations, both of these extremes are possible, as are all possibilities in between. The end result is that if spontaneous mutations are responsible for the appearance of the phage-resistance in bacteria, then one should find significant fluctuations in the numbers of resistant cells from tube to tube. If, on the other hand, phage resistance is due to some chance adaptation mechanism, in which every cell has an equal but small probability of becoming resistant, then there should be no fluctuations in the numbers of resistant cells among the tubes.

TABLE XI
REPRESENTATIVE DATA FROM THE LURIA-DELBRÜCK EXPERIMENT

Separate cultures		Single culture	
Tube Number	Phage-resistant colonies	Tube Number	Phage-resistant colonies
1	30	1	46
2	10	2	56
3	40	3	52
4	45	4	48
5	183	5	65
6	12	6	44
7	173	7	49
8	23	8	51
9	57	9	56
10	51	10	47
mean = 62		mean = 51.4	
variance(corrected)* = 3498		variance(corrected)* = 27	

*Because they did not sample the entire tube contents, Luria and Delbrück applied a small sampling correction to their calculations of variance. If the variances were calculated directly from equation in the text, they would be 3999 and 39 for the separate cultures and the single culture, respectively. Data from Luria, S. E., and Delbrück, M.: *Genetics, 28:*491, 1943.

The results of a Luria-Delbrück experiment are shown in Table XI. In the left-hand column are shown the numbers of resistant colonies appearing on the phage plates. It would appear that the numbers do in fact fluctuate. The key is whether the fluctuations are significant in a statistical sense, and are not due to some outside influence such as pipetting error. To test the level of fluctuations due to sampling error, the same culture tube was repeatedly sampled, and each aliquot was spread onto a phage plate. The results of this test are shown in the right-hand column of Table XI. These counts also show fluctuations, but to compare the fluctuations from each series of plates, Luria and Delbrück resorted to a statistical parameter known as variance. Variance is a measure of how widely values in a series fluctuate about the mean. Variance is calculated by finding the difference between each value (n) and the mean (ñ) (known as the deviation), squaring the deviations, adding them up, and dividing by the number of values (t) minus one. In mathematical notation this would be:

$$\text{Variance} = \frac{\Sigma(n-\bar{n})^2}{t-1}$$

The variance for the single tube comes out to be 27, whereas the variance for the ten individual tubes is 3498. Thus, the degree of fluctuations among the individual tubes was considerably greater than that from the single tube. This strongly supported the hypothesis that the bacteria had acquired bacteriophage resistance as a result of spontaneous mutations.

Newcombe Spreading Experiment

The results of the Luria-Delbrück experiment were not widely accepted by microbiologists. Critics pointed to the fact that it was not proved that the phage on the agar plates did not induce resistance in the bacteria. Besides, microbiologists were not generally accustomed to using and understanding statistics. Six years passed, and another bacteriologist, H. B. Newcombe, devised an experiment in 1949 which involved the plating of susceptible bacteria onto agar plates (Fig. 42). Following a short incubation period, half the plates were respread with a sterile

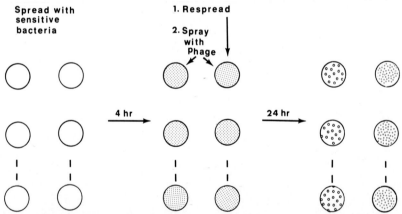

Figure 42. The Newcombe experiment. A number of nutrient agar plates was spread with phage-sensitive bacteria and incubated for 4 hours. One half of the plates was respread with sterile stirring rods, after which all of the plates were sprayed with phage and reincubated for 24 hours. The numbers of colonies appearing on the respread plates compared with those on the other plates supported the mutation origin of the resistant bacteria. There were more colonies on the respread plates than on the others. If adaptation had been responsible for phage resistance, the same number of colonies should have appeared on all plates.

stirring rod, and all were then sprayed with a suspension of bacteriophage. Following incubation, the respread plates had obviously more resistant colonies growing than the plates not respread. Newcombe reasoned that all plates had the same number of cells on them, as each was inoculated with an identical quantity of bacteria, and all were incubated under identical conditions. Thus, if phage resistance was acquired by adaptation, the same number of resistant colonies should develop on every plate. That the respread plates contained many more colonies supported the hypothesis of the mutational origin of bacteriophage resistance.

Lederberg Indirect Selection Experiment

Newcombe's experiment convinced a few more bacteriologists, but it remained for Joshua and Esther Lederberg to devise a test that avoided exposing the bacteria directly to the phage. You recall that the criticism of the Luria-Delbrück experiment, as well as that of Newcombe, revolved about the question of whether the presence of the bacteriophage may be inducing the changes to resistance, rather than the resistance being due to undirected, spontaneous mutations. In the words of the Lederbergs:

"Elective enrichment is an indispensable technique in bacterial physiology and genetics.* Specific biotypes are most readily isolated by the establishment of cultural conditions that favor their growth or survival. It has been repeatedly questioned, however, whether a selective environment may not only select but also direct adaptive heritable changes."

The Lederbergs' experiment was based on a technique of indirect selection, which allows for the isolation of bacteria exhibiting a given phenotype, such as phage-resistance, without exposing the cells to the selective agent, the *phage*. The technique, also known as *replica plating,* has since proved to be a valuable tool in many applications in the microbial laboratory (Fig. 43).

---

*In quoting C. B. Van Niel

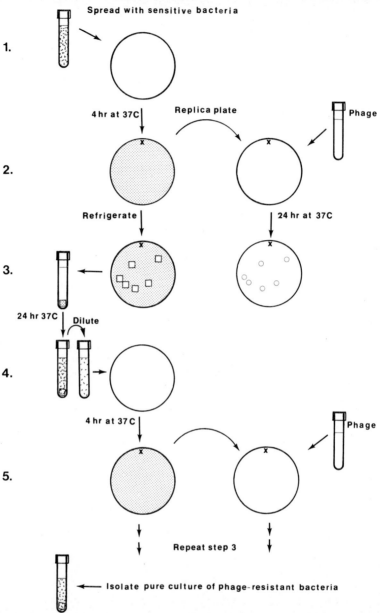

1.
Spread with sensitive bacteria

2.
4 hr at 37C | Replica plate | Phage

3.
Refrigerate | 24 hr at 37C

4.
24 hr 37C | Dilute

5.
4 hr at 37 C | Phage

Repeat step 3

Isolate pure culture of phage-resistant bacteria

Figure 43. The Lederberg Indirect Selection Experiment. Phage-sensitive bacteria were spread onto a nutrient agar plate and the plate was incubated for 4 hours. A sterile velveteen pad was pressed against the sparse bacterial growth on the plate and then against the surface of a fresh plate that had been seeded with bacteriophages. The seeded plate was incubated for 24 hours; the original plate was refrigerated. On the following day (step 3), a few colonies of phage-resistant bacteria were observed on the plate that had been incubated. The locations of the original phage-resistant clones on the refrigerated plate were determined. The agar at those locations was excised aseptically and transferred to tubes of nutrient broth, which were incubated. Steps 1, 2, and 3 were repeated until a pure culture of phage-resistant bacteria was isolated. The bacteria in such a culture had never been exposed to the phage; they had been selected for indirectly.

Approximately 1 to 5 x 10^7 phage-sensitive bacteria were spread onto the surface of a nutrient agar plate and subjected to incubation for four to six hours. Following the incubation period, a piece of sterile velveteen cloth was pressed against the agar surface and then against the surface of a fresh agar plate that had been seeded with phage. The nap of the velveteen simulated thousands of inoculating needles and thus transferred a fraction of the growth to the fresh plate. In addition, the velveteen maintained the relative juxtapositions of the members of each original clone on the phage plate. The phage plate was incubated overnight; the original, or master plate was refrigerated.

On the following day, examination of the phage plate revealed the development of a few phage-resistant colonies, the majority of the cells transferred being of course lysed by the phage. By referring back to the original plate, one determined the approximate location of the clone from whence the resistant colony on the phage plate originated. This was aided by marking the original and phage plates when the replica plating was performed. Using a sterile spatula, one then removed the small block of agar from the master plate that may contain the parent clone. The block was dropped into a tube of nutrient broth and incubation was carried out for twenty-four hours.

The resulting growth was diluted appropriately and then spread on an agar surface and subjected to another four to six hours of incubation. Replica plating of the resulting growth was again carried out onto phage plates, repeating the entire cycle as described two to three more times. Finally, in the last cycle, the master plate was incubated overnight rather than refrigerated. Discrete colonies appear on the master plate that, when tested, all appeared to be resistant to bacteriophage attack. It should be remembered that these bacteria on the master plates were never exposed to bacteriophage, only those on the replica plates, and yet one has been able to select for several phage-resistant variants from the original susceptible culture. This is what is meant by *indirect* selection. The phage-resistant variants arose through the occurrence of spontaneous mutations, and the technique of replica plating made it possible to isolate the mutants without their exposure to phage.

The Lederberg experiment just described is frequently thought of as the capstone of the theory of the mutational origin of variants in bacteria.

MOLECULAR BIOLOGY OF MUTATIONS

Mutations, as we defined them in Chapter One, are changes in the genotype of an organism. Genotypic information is encoded in the nucleotide sequences of the nucleic acid making up the chromosome of the organism, and occasionally an extrachromasomal piece of nucleic acid as well. Hence, we can consider mutations as alterations in this sequence. These alterations can be categorized into basic types: If a purine is replaced by another purine, the mutation is referred to as a *transition*. A *transversion* is a mutation in which a purine is replaced by a pyrimidine, or vice-versa. We have already alluded to *deletions* and *additions* (or *insertions*) in our discussion of the genetic code. A deletion is the removal of one or more nucleotides from the sequence in a gene; an addition is the placement of one or more extra nucleotides within the gene. You recall that when a deletion or additon involved a small number of nucleotides, other than three or multiples of three, a *frame-shift* mutation occurred in which the triplet translation reading frame is out of phase. Transitions, transversions and other mutations that involve only one nucleotide are frequently called *point mutations*. Deletions and additions may span two or more genes and involve 1000 to 2000 nucleotides; these are sometimes referred to as *macrolesions*. Another example of a macrolesion is a *duplication,* in which a large segment of the genotype is duplicated in the nucleic acid chain. (It is interesting that the genotypes of some bacteriophages and many higher plants and animals appear to contain natural duplications. It is thought that duplications have played a significant role in the evolution of genes.)

Rearrangements are another type of macrolesion that involve a large segment of a gene being either deleted and reinserted in another location on the chromosome (or perhaps on another chromosome, in eucarytic organisms), or deleted and reinserted in the same location, but inverted. The latter case is known as an *inversion*.

Microorganisms are known to experience all of the different types of mutations cited above. They may occur either spontaneously, or be induced by any of a number of physical and chemical agents known as *mutagens* that may be applied to a culture in the laboratory. Examples of these various types of mutations are diagrammed in Figure 44.

Spontaneous Mutations

All cells (including viruses) experience rare, randomly-occurring spontaneous mutations. For a given mutation, such as the acquisition of penicillin resistance in bacteria, or hemophilia in humans, the rate of its natural occurrence can be estimated and is relatively constant. But what is the cause of spontaneous mutations? Do they "just happen," or are there

Wild Gene Sequence   GATGATGATGAT......

Transition (pu→pu or py→py)   GATGA*CGATGAT......

Transversion (pu⇌py)   GATGAT*TATGAT.....

Deletion   *GTGATGATGAT.....

Insertion   GAT*CGATGATGAT......

Inversion   GAT*TA*GGATGAT.......

Figure 44. Examples of types of mutation. Asterisk indicates location of mutation. Abbreviations: Pu = purine; py = pyrimidine.

external forces that bring them about? There are probably several causes of spontaneous mutations; more significant ones are discussed below:

1. *Cosmic and Background Radiation*—We have referred to Muller's discovery (1927) that X rays will induce mutations in fruit fly larvae. This work led to the supposition that cosmic radiation and natural background radioactivity, to which every living organism on earth is exposed, are the main causes of spontaneous mutations. However, by comparing various radiation intensitites with the number of observed mutations and extrapolating the radiation intensity to background level, others could show that the intensity of natural ionizing radiation on the earth's surface accounted for only a small fraction of spontaneous mutations observed (Fig. 45). Thus, while background radiation may cause some spontaneous mutations, other factors appear to contribute to their incidence.

2. *Ultraviolet Light*—Ultraviolet light, usually considered that portion of the light spectrum between 100 and 400 nm (nanometers), is found abundantly in solar radiation, although the earth's atmosphere filters out most of it. It has been found that UV radiation around 260 nm is highly mutagenic to microorganisms, small insect larvae, and other organisms small enough for the radiation to penetrate, for the 260 nm band cannot pass through more than a few micrometers of tissue. Exposure to UV light is not confined to those organisms exposed to strong sunlight, for the use of germicidal UV lamps in hospitals and laboratories is now widespread. It is not known how significant a role ambient UV radiation plays in the formation of spontaneous mutations in microorganisms, but it is hypothesized that it played a significant role in the evolution of life on earth some three billion years ago when all life was single-celled and the UV portion of the sun's radiation was considerably more intense at the earth's surface.

3. *Mutagenic Chemicals*—Over ten years after Muller's findings, it was shown that certain chemical compounds also can induce mutations in organisms. This early work involved highly toxic materials used in chemical warfare that are not normally found in the environment. However, since that time (1940's)

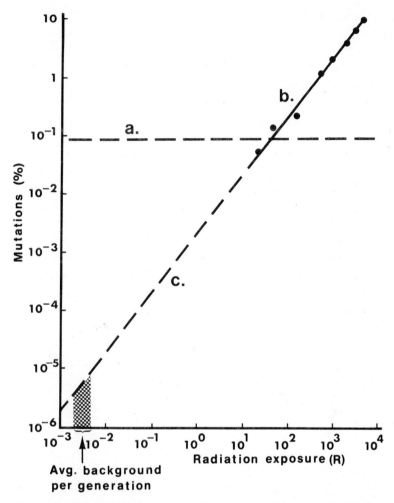

Figure 45. Role of background radiation in origin of spontaneous mutations. The spontaneous rate for sex-linked recessive lethal mutations in *Drosophila* is about .1% (a.). If the same mutations are induced by x-radiation, the observed mutation rates, corrected for the spontaneous rate, fall about the straight line (b.). This line is then extrapolated to the level of normal background radiation received by the flies per generation (c.). One could then theorize that the mutation rate due to the background radiation would be of the order of 10^{-6}%. The true spontaneous rate is nearly five orders of magnitude higher, however. (Radiation exposure is in roentgen units (R)). (From Von Borstel, R. C.: *Japan J Genetics, 44,* Supp. 1:102, 1969)

many common, seemingly innocuous compounds have been found under certain conditions to induce mutations in microorganisms and other test species. Included in these compounds are many familiar food additives, drugs, and pesticides. It is not unreasonable to hypothesize that some spontaneous mutations may be induced by such compounds that have accumulated in the environment and from there gain access to the genetic material of organisms.

4. *Errors in DNA Replication and Repair*—There is growing evidence that most spontaneous mutations owe their existence to errors in DNA replication and repair. This is suggested by experiments with bacteria and bacteriophages that have experienced *mutator* mutations. These mutations are characterized by an overall increase in the rates at which spontaneous mutations occur in the organisms. Careful analysis of the mutant cells reveals that they have suffered some alteration in the cell's machinery responsible for DNA replication or repair. This aspect will be covered in more detail in the section on mutator genes.

INDUCED MUTATIONS

The discovery of the mutagenic power of X rays in 1927 created considerable excitement among geneticists, for mutations are the coin of genetics research, and to be able to increase their numbers at will by a hundred-fold or more was a significant advance. The excitement was somewhat modulated however, by the realization that the artificial induction of mutations by X rays and chemicals is nearly as undirected and random as is the occurrence of spontaneous mutations. Nevertheless, the use of mutagens to induce mutations in organisms in the laboratory has been of inestimable value. Let us now review some of the agents used in mutagenesis.

Physical Mutagens

Ionizing Radiations

Figure 46 depicts the spectrum of electromagnetic energy, showing the relationships between various types of radiation.

In the region of greater energies are X rays, discovered by the German physicist Röntgen in 1895. Because of their great power of penetration, X rays have been very useful in the deliberate induction of mutations in multicellular plants and animals since 1927. However, in the case of microorganisms, while some laboratories may use X rays for inducing mutations, the use of this form of energy has not been widespread. The initial cost of X-ray equipment is high, as is its maintenance, and there is a certain degree of risk involved to the operators of such equipment. Furthermore, the effects of X rays on cells are complex and little understood. Unless one is specifically interested in the effects of ionizing radiation on organisms, there is little reason to use this type of radiation for the routine induction of mutations in microorganisms.

Commercial units that deliver beta or gamma radiation from radioactive isotopes such as cobalt 60 or cesium 137 are more convenient to use than X-ray sources. The isotope is usually sealed in a massive lead shield, and either the sample to be irradiated is lowered into a well or the source is raised through a series of elevators and baffles until the two are in close proximity. A typical unit is shown in Figure 47.

The mechanism of mutagenesis by ionizing radiation is not fully known. Chromosome breaks and other microscopically visible consequences of the radiation are most frequently cited. Ionizing radiation produces what are known as indirect effects as well as direct effects. That is, the radiation may activate compounds in the medium or cytoplasm that in turn bring about mutations at the same time the radiation is directly damaging the DNA of the chromosome.

Ultraviolet Light

In contrast to ionizing types of radiation, ultraviolet light is a practical, inexpensive, and effective means of inducing mutations in microorganisms. It was not widely used for this purpose until the work of Demerec and Laterjet in 1946, even though the mutagenic power of ultraviolet radiation had been known

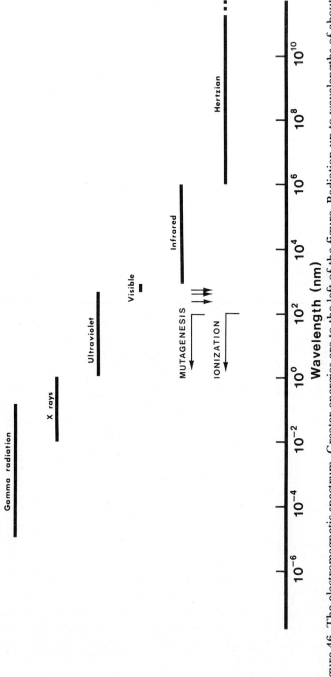

Figure 46. The electromagnetic spectrum. Greater energies are to the left of the figure. Radiation up to wavelengths of about 10^{-2} nanometers are sufficient to bring about ionization of molecules and mutagenesis in cells. Mutagenicity is also observed in the narrow region of maximum light absorption of the nucleic acids, 263 nm, and at certain wavelengths in the visible region. Hertzian radiation includes radio and radar frequencies.

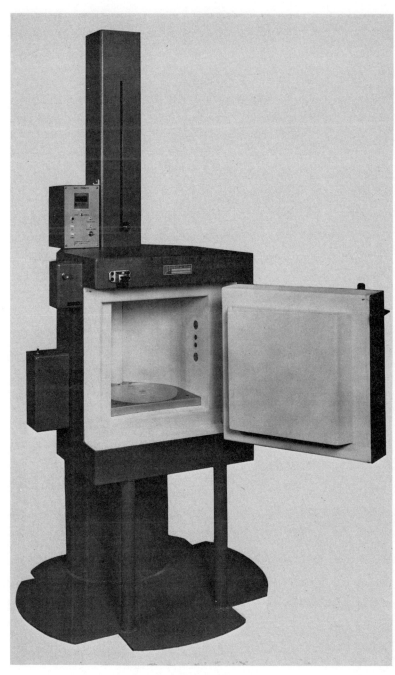

Figure 47. Example of a cesium 137 irradiator. Objects are placed in the center chamber, the door is closed and the ^{137}Cs source is raised into position to expose the objects to an intense flux of gamma irradiation. This unit stands 6 feet tall and weighs 6000 lbs. (Photo courtesy J. L. Shepherd and Associates, Glendale, California.)

since the early 1930's. The most convenient sources of mutagenic UV light for laboratory use are the germicidal lamps manufactured by General Electric, Westinghouse, and others. Care must be taken in selecting a lamp for this purpose, for there are UV lamps sold for other applications that are not satisfactory, such as those for the examination of minerals (Wood's lamps), or for the illumination of fluorescent posters and other displays (black lights). These lamps do not emit sufficient radiation at the mutagenic wavelengths, around 260 nm, to make them useful for inducing mutations. True germicidal lamps are usually of the low-pressure mercury vapor type. The emission spectrum of mercury under these conditions includes an intense line at 253.7 nm, which is within the mutagenic and lethal region of the UV spectrum (Fig. 48).

This last statement alludes to the fact that UV light really has two effects on organisms, it can induce mutations and at the

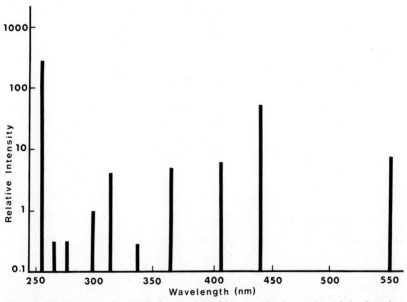

Figure 48. Output of a typical germicidal lamp. The intensity of the band at 253.7 is over 100 times greater than the average intensity of all the other major bands up to 550 nm.

same time it can also kill them. (This distinction is only an operational one, for mutagenicity and lethality may share the same mechanism.) The interrelationship of the two effects is shown in Figure 49. When a suspension of microorganisms is irradiated with a source of UV light of around 260 nm, the cells are killed in a logarithmic fashion. That is, for each equal dose of radiation the same fraction of cells is killed. (See also Fig. 76).

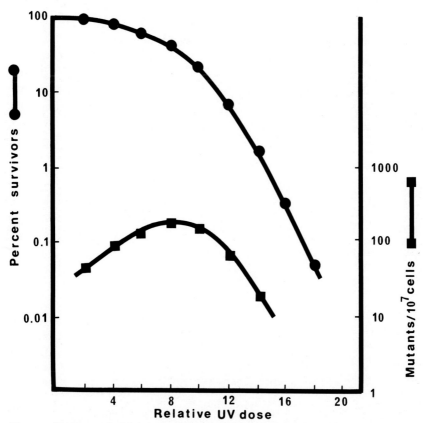

Figure 49. Bacteriocidal and mutagenic characteristics of ultraviolet light. Bacteria exposed to radiation at about 260 nm die off at an exponential rate (circles). As dose is increased, the number of mutants isolated (squares) increases to a point, then drops off. In this instance, the optimum dose for inducing mutations would be that which corresponds to a survival level of about 50%. (Data from Hill, R., *Photochem Photobiol, 4:*563, 1965)

If at the same time one is assaying the suspension for viable cells, the number of mutations induced as a function of dose is also observed, one finds that the number of mutations rises to a peak and then drops off (Fig. 49). It would appear that at low doses the number of mutations induced is proportional to the dose absorbed by the cells, but there is a dose level above which more cells are being killed than there are mutants surviving. The UV dose at which maximal mutations are induced must be determined experimentally for each strain of organism, for it can vary from a dose corresponding to a survival level of 50 percent down to one of 0.1 percent or less.

It generally has been found that the principal damage incurred by mutagenic UV light is the formation of dimers between adjacent pyrimidine nucleotides in the DNA molecule, the most common being between two thymine molecules (Fig. 50). The carbon-carbon bonds forming the dimers are somewhat shorter than the normal distance between adjacent bases, resulting in a deformation of the DNA chain (Fig. 51).

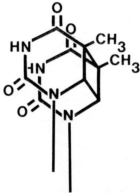

Figure 50. A dimer formed of two molecules of thymine as a result of ultraviolet radiation. Covalent bonds are formed at the number 4 and 5 carbon atoms, binding the two pyrimidines together.

The exact nature of the dimer's role in mutagenesis is not clear, but it can be hypothesized that errors in replication or repair may be involved in which either the wrong base is inserted in position on the opposite DNA strand, an extra base may be added or one may be deleted.

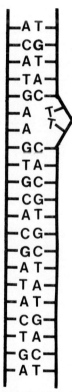

Figure 51. Effect of a thymine-thymine dimer in DNA. A portion of double-stranded DNA is diagrammatically shown here containing a thymine-thymine dimer, which causes a distortion of the normal configuration of the molecule.

Chemical Mutagens

Just before and during the World War II, considerable research on chemical warfare was carried out in the British

Isles. One of the classes of compounds most intensively studied was the nitrogen mustards. These are chemicals related to the mustard gases used in the World War I in which a nitrogen has replaced the sulfur (Fig. 52). All of these compounds are highly toxic, bringing about very severe and painful tissue burns in animals that heal only slowly. These are also the characteristics of burns brought about by ionizing radiation, and for that reason the mustard compounds and other similar materials are frequently referred to as being *radiomimetic*. Since ionizing radiation had been shown to induce mutations in organisms, it was tested whether radiomimetic compounds did so as well.

$$\text{ClCH}_2\text{-CH}_2\text{-NH-CH}_2\text{-CH}_2\text{Cl}$$

Figure 52. Example of one of the nitrogen mustards. This is di-(2-chloroethyl)amine.

Thus, it was discovered that the nitrogen mustards are excellent mutagens, and after much research hundreds of other compounds have been found to induce mutations in organisms. At first it was thought that a compound had to be radiomimetic to be mutagenic, but it has subsequently been demonstrated that a compound need *not* be radiomimetic to be mutagenic, nor does it need necessarily to be very toxic. A number of only mildly toxic chemicals, such as caffeine and nitrous acid (the sodium salt of which is a common food preservative) are highly mutagenic for microorganisms. We will now review some of the more commonly used mutagenic chemicals.

Nitrous Acid

HNO_2 is commonly used in inducing mutations in viruses and in bacteria. The most obvious action of this agent is the oxidative deamination of guanine, adenine, and cytosine, converting them to xanthine, hypoxanthine, and uracil, respectively (Fig. 53). Xanthine appears to have about the same base-pairing characteristics as guanine, so no mutation would be expected here, but hypoxanthine more readily base-pairs with cytosine, leading to the eventual replacement of the original A:T pair with a G:C pair. Uracil pairs with adenine, resulting in a C:G to a T:A transition. In addition to the obvious point mutations possible with nitrous acid, deletions and other changes in DNA have also been reported, but their mechanisms have not been explained.

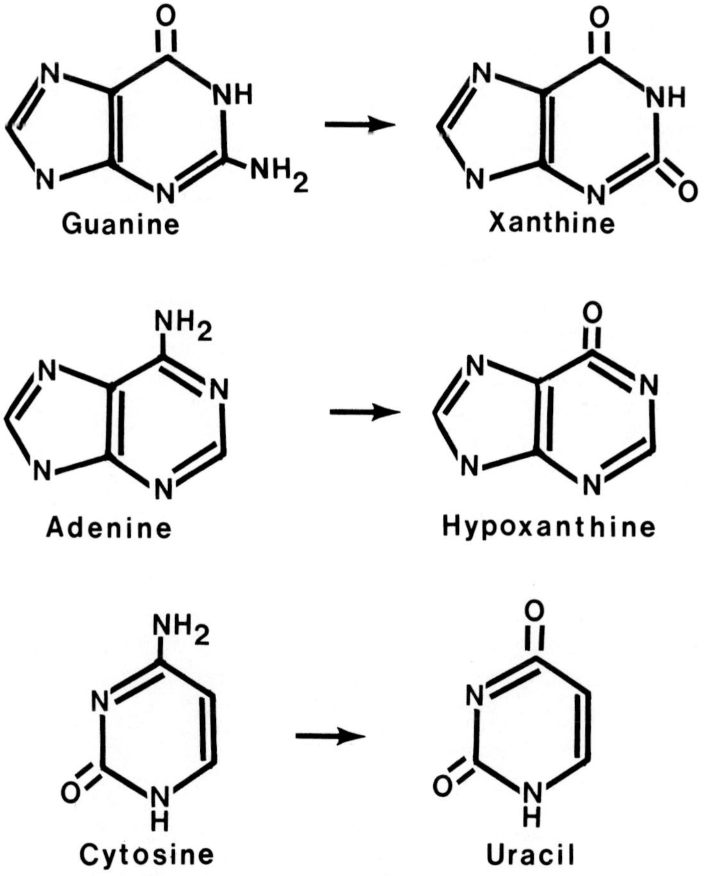

Figure 53. The action of nitrous acid on purine and pyrimidine bases.

Hydroxylamine

NH_2OH is most effective on free virus particles or pure DNA. Its most striking action is the formation of a derivative of cytosine, 4,5-dihydro-4-hydroxylaminocytosine, from cytosine (Fig. 54). The derivative has altered base-pairing tendency, preferring to pair with adenine, and thus resulting in the transition C:G to T:A.

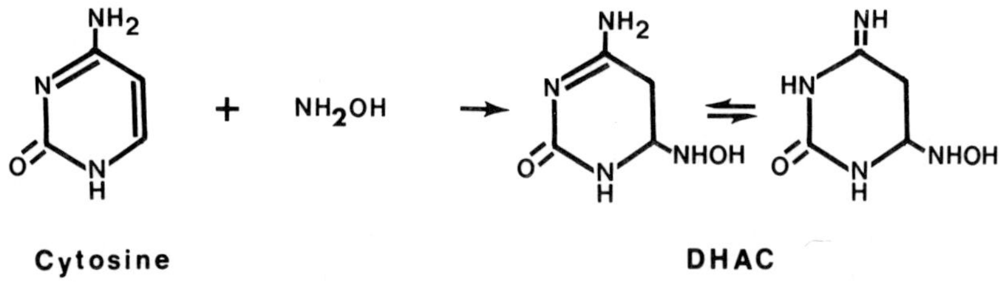

Cytosine   DHAC

Figure 54. The mutagenic action of hydroxylamine is focused principally on cytosine, which is oxidatively aminated to 4,5 dihydro-4-hydroxylaminocytosine (DHAC). Two tautomers of DHAC are possible, both of which appear to base pair with adenine rather than with guanine.

Nitrogen Mustards

These compounds were the first chemical mutagens to be discovered, and are effective for all microorganisms, as well as higher plants and animals. They belong to a general class of chemical compounds known as alkylating agents, which are known to add alkyl side chains to the DNA molecule, particularly to the 7 position of guanine and the 3 position of adenine. In the case of the mustard compounds, which are polyfunctional (having more than one reactive group), covalent bridges may form between adjacent bases or between opposite strands of the DNA.

Other examples of alkylating agents are ethylmethane sulfonate, ethylethane sulfonate, and nitrosoguanidine.

Nitrosoguanidine

This compound, whose full chemical name is N-methyl-N'-nitro-N-nitrosoguanidine (NTG), is one of the most effective mutagens known (Fig. 55). Under proper conditions, a treated culture of bacteria may exhibit over 50 percent mutated cells when exposed to NTG. The most common result of NTG treatment appears to be G:C to A:T transitions, although some transversions have also been observed. It also is highly effective in inducing mutations in bacteriophage particles but only when applied to the infected bacterial hosts.

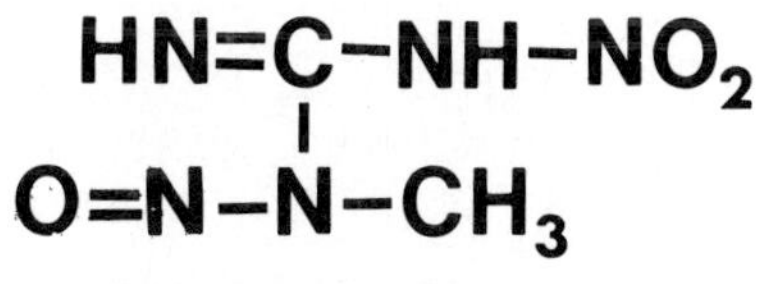

$$HN{=}\underset{|}{C}{-}NH{-}NO_2$$
$$O{=}N{-}N{-}CH_3$$

Nitrosoguanidine
Figure 55. The structure of ni-trosoguanidine (NTG).

The action of NTG appears to be centered at the DNA replication fork. It has been demonstrated that NTG mutagenesis can be directed at specific regions of the chromosome by synchronizing the DNA replication cycle and then exposing the cells to NTG at the moment the replication fork is passing the target region. Locations of specific genes can thus be determined.

Since both transitions and transversions are observed with other alkylating mutagens, the altered bases appear to induce more than just errors in base pairing. It has been proposed that the alkylation of the 7 position of guanine weakens its bonding with deoxyribose, leading to an excision of the purine. On DNA replication, the resulting gap may then be copied on the opposite strand with any random purine or pyrimidine. Cross-linkages between adjacent bases formed by polyfunctional alkylating mutagens may produce replication errors similar to those in UV-induced dimers. Cross-links between opposite DNA strands prevent complete strand separation, but one can only hypothesize how these may bring about mutations.

Although all mutagens must be handled with great care, the alkylating agents demand particular attention. They appear to persist in the environment for long periods, and their toxicities are not completely known. There is evidence that NTG may be carcinogenic.

5-Bromouracil

5-BU, in which a bromine atom has replaced the methyl group of thymine, is one of a few mutagens known as *base analogs*. A base analog is a compound that resembles a natural base sufficiently that the analog may be incorporated in the DNA strand in place of the natural base. 5-BU normally base pairs with adenine, as does thymine, through the keto oxygen at position 6, and the nitrogen at position 1. However, as a consequence of what is known as a tautomeric shift in which the hydrogen of the amino group at position 1 may infrequently shift to the keto oxygen, the 5-BU now has a greater affinity for guanine. The end result could be an A:T to a G:C transition (Fig. 56). Other changes also are noted with 5-BU, leading us to believe that we do not have the full story concerning the action of this mutagen.

In order to be effective, a base analog must be incorporated into the DNA during its replication. Making the base analog available in the medium frequently is not sufficient for incorporation, however. Many organisms, such as *E. coli*, synthesize their thymine needs *de novo* and the natural pyrimidine competes very successfully with 5-BU for incorporation. If the cells are either genetically or chemically prevented from making thymine, however, the cells are forced to take up the 5-BU into the DNA.

The most common technique to force-feed a culture of bacteria with 5-BU is to grow them in the presence of sulfanilamide. This drug blocks the formation of folic acid, a key cofactor in the anabolism of several metabolites, including thymine. All of the blocked metabolites are supplied in the

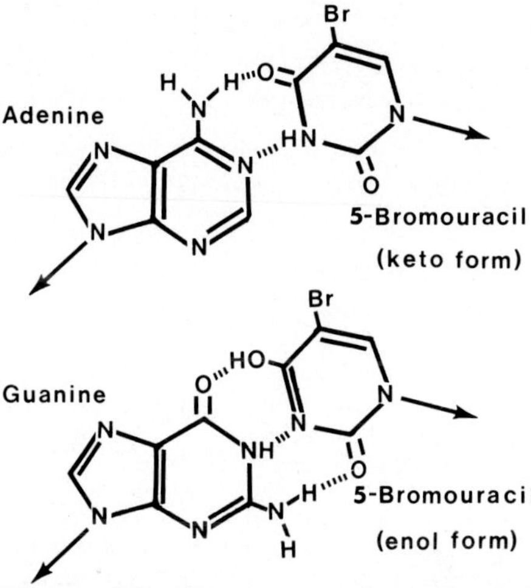

Figure 56. The mutagenic action of 5-bromouracil (5BU). The more common tautomer of 5BU normally base pairs with adenine (upper figure) as does thymine for which it is an analog. A rarer enol form (lower figure) more readily base pairs with guanine, leading to an AT to GC transition. Arrows indicate points of attachment to deoxyribose.

medium except for thymine, which is replaced by 5-BU. By this technique all of the thymine residues of the cells' DNA eventually are replaced by the analog. 5-BU apparently is not a perfect substitute, however, for in addition to the rare mutations it may induce, it leaves the cells considerably more susceptible to the lethal actions of many physical and chemical agents.

Base analog incorporation into bacteriophage DNA is carried out in a similar manner in that the host bacteria are treated as above.

Acridines

The acridine dyes make up a large class of organic compounds characterized by molecular structures of three fused rings (Fig. 57). The most effective mutagens of the group are proflavin, acriflavin, and acridine yellow. While it is clear that acridine treatment brings about various types of mutations in microorganisms, predominantly frame shift, it is not at all clear what mechanisms are involved. It is theorized that the nearly flat acridine molecules intercalate between nucleotides in the DNA chain, distorting the molecule and thereby inducing errors during replication, repair, or possibly recombination.

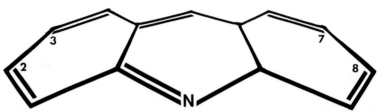

Figure 57. The general structure of the acridines, drawn to emphasize the flat nature of the molecule. Some of the common acridines used as mutagens are: proflavin, (2, 8-diaminoacridine), acridine orange, (2, 8-bisdimethylaminoacridine), and acridine yellow, (2, 8-diamino-3, 7-dimethylacridine).

As with the base analogs, the acridine mutagens must be applied during DNA replication. Not all microorganisms are equally susceptible to acridine mutagenesis. For example, proflavin is highly mutagenic towards the T-phages but not against phage lambda. It is not particularly effective for *E. coli* except when the cells undergo conjugation.

To complicate the use of the acridine dyes further, these compounds are known to participate in a phenomenon known as *photodynamic action*. In the presence of bright light, incorporated acridine compounds will induce additional mutations of the transition and transversion types. The mechanism of this action is not fully known, but since only light of certain wavelengths is effective, presumably the dyes absorb some of the light energy and rise to a hyperactive state. The energy then is transferred to the DNA to induce damage that later leads to a mutation.

ICR Compounds

A number of compounds have been synthesized by chemists at the Institute for Cancer Research that are effective mutagens. These materials are based on the acridine molecule, but with additional, often alkylating side-chains. An example of one of the ICR mutagens is ICR-191 (Fig. 58). The most prominent action of these compounds appears to be the induction of frame-shift mutations.

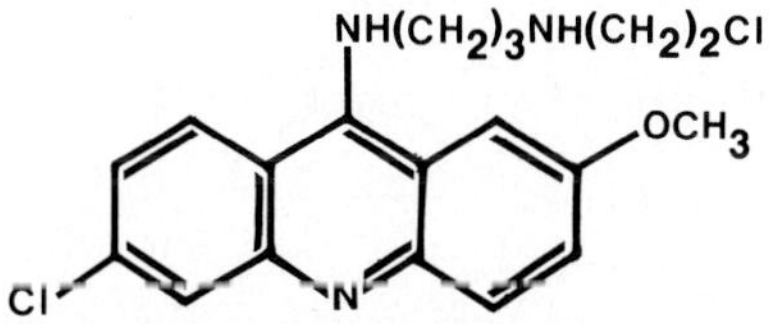

Figure 58. Structure of one of the ICR mutagens. This one is ICR-191, a relatively strong mutagen of the series. (From Ames, B. N. and Whitfield, H. J., Jr.: *Cold Spring Harbor Symp Quant Biol, 31*:221, 1966).

Miscellaneous Chemical Mutagens

Quite a number of other chemical compounds have been shown to induce mutations in microorganisms to one degree or another. Examples are the pesticide captan, the antibiotics streptomycin and mitomycin C, caffeine, and inorganic salts such as $MnCl_2$ and $FeCl_2$. With the exception of mitomycin C, none of these compounds has been favored with wide use. Caffeine has attracted some attention in that it appears that its mutagenic action stems from its interference with cells' normal DNA repair mechanisms.

In spite of our apparent knowledge of the actions of many mutagens, the exact mechanisms whereby they induce mutations are not at all clear. A number of perplexing factors seem to play a role, as for example, the nature of neighboring bases. There is information to suggest that the creation of a mutational lesion frequently depends on the nature of the nucleotide bases immediately adjacent to the base directly affected by the mutagen.

MUTATION RATES

Methods of Estimation

The strict definition of mutation rate considers it as a probability that a single bacterial cell will experience a specific mutation during a division cycle. For these purposes a division cycle is that interval of time between consecutive divisions of the bacterial cell. Thus mutation rate (a) can be expressed:

$$a = \frac{M}{D} \qquad M = \text{mutations} \qquad (1)$$

$$D = \text{division cycles}$$

If the initial number of cells is small, the number of division cycles (D) experienced by a culture is very nearly equal to the total number of bacteria (N). Thus $D = N$, and by rearranging equation (1) we get

$$M = aN \qquad (2)$$

That is, the number of mutations equals the mutation rate times the number of cells.

Luria and Delbrück posited two methods for estimating mutation rates, one of which is based on a statistical concept devised by the French mathematician Simeon Poisson (1781-1840). The concept, known as Poisson's Distribution, states that for x random events, given y opportunities to occur, the probability (P) that z events will occur per opportunity is

$$P(z) = \frac{(x/y)^z \, e^{-(x/y)}}{z!} \qquad (3)$$

Luria and Delbrück saw the random events as spontaneous mutations, the opportunities as culture tubes, and z as the average number of mutations occurring per tube.

These workers reduced the Poisson equation so that z = 0, that is, the case in which no mutations occurred. In doing so, equation (3) simplifies to

$$P(0) = e^{-(x/y)} \qquad (4)$$

Since x/y equals mutations, the same as M in equation (2), we can substitute aN for x/y in equation (4), and rearrange it to arrive at

$$a = -\frac{\ln P(0)}{N} \qquad (5)$$

The mutation rate (a) is thus estimated by dividing the natural logarithm of the proportion of tubes that contained no mutations by the average number of cells per tube. This estimation is in terms of mutations per cell division.

It is more common to express mutation rates in terms of mutations per cell per division, in which case equation (5) is amended by substituting N/ln 2 for the N. The expression, N/ln 2, is a more accurate estimation of the actual number of cells per tube. The final equation for the estimation of mutation rates is then:

$$a = -\frac{\ln P(0)}{N/\ln 2} \qquad (6)$$

This method is most accurate when the experiment is adjusted such that P(0), the proportion of tubes showing no mutations, is roughly 1/3. A certain amount of trial and error is involved, in which inoculum size, and culture volume, are manipulated to arrive at this proportion. Table XII lists data from an experiment of Luria and Delbrück in which conditions were adjusted (quite successfully) to yield P(0) = 1/3.

TABLE XII
MUTATION RATE ESTIMATION FROM RESULTS OF THE LURIA-DELBRÜCK EXPERIMENT*

Total number of tubes	87
Tubes with no mutants:	29
Average population per tube:	2.4×10^8
Mutation rate (by equation 6):	0.32×10^{-8}
Mutation rate (by equation 7):	2.4×10^{-8}

*Data from Luria, S. E., and Delbrück, M.: Genetics, 28:491, 1943.

The application of equation (6) for estimating mutation rates has both good and bad points. Because one is considering only those tubes in which *no* mutations occurred, one need not be concerned with differential growth rates of parent and mutant cells, a source of error by other methods. However, the method requires large numbers of tubes (at least 100) to be effective and conditions requiring difficult adjustments to conform to the 1/3 rule. The entire volume of each tube must be assayed. Further, the method really does not make full use of the data available.

The second method of Luria and Delbrück utilizes more fully the data generated by a fluctuation experiment. The method is based on a somewhat more complex formula:

$$r = aN_t \ln (aN_t C) \qquad (7)$$

where r = average number of mutants per sample
a = mutation rate, mutations per cell per generation
N_t = average number of bacteria per tube (or sample)
C = number of culture tubes (or samples)

Equation (7) cannot be rearranged easily to calculate the value of a directly. Experimentally derived values of r, N_t and C are substituted into the equation and the value of a is solved for algebraically by utilizing graphs in which r is plotted against aN_t for various values of C. Such graphs can be found in Luria and Delbrück's original paper, or in Clowes and Hayes, *Experiments in Microbial Genetics.*

The last method we will describe was devised by Newcombe. A series of four plates is inoculated with identical quantities of appropriate bacteria and incubated. At some later time two plates are removed, one is sprayed with the selective agent (bacteriophage, antibiotic, etc.) and reincubated. The total bacterial population is determined on the second plate by washing the cells off of it with sterile saline and assaying the saline by standard viable plate count methods. At yet a later time, the second pair of plates is removed from incubation and treated as above. The rate for the observed mutation can be estimated by applying the formula:

$$a = \frac{(M_2 - M_1)}{(N_2 - N_1)/\ln 2} \qquad (8)$$

M₁ and M₂ are the number of mutant colonies appearing after times 1 and 2, respectively. N₁ and N₂ are the numbers of cells on the assay plates at times 1 and 2. For best results, N₂ should not exceed about 10^9 cells, and times 1 and 2 should be well within the midlog growth phase of the bacteria being studied, i.e. for *E. coli,* six to eight hours or less.

Table XIII lists mutation rates estimated for a number of phenotypes in several species of bacteria. All but one rate were determined by Luria and Delbrück's first method (equation (6)).

TABLE XIII
EXAMPLES OF MUTATION RATES IN BACTERIA*

Mutation	*Species*	*Mutation rate per cell per generation*
Pigmentation loss	*Serratia marcescens*	1×10^{-4}
Radiation resistance	*Escherichia coli*	1×10^{-5}
Isoniazide resistance	*Mycobact. ranae*	3×10^{-6}
Penicillin resistance	*Staphylococcus aureus*	1×10^{-7}
T3 bacteriophage resistance	*E. coli*	1×10^{-7}
T1 bacteriophage resistance	*E. coli*	3×10^{-8}
Sulfathiozole resistance	*Staphylococcus aureus*	1×10^{-9}
Loss of galactose fermentation	*E. coli*	1×10^{-10}
Streptomycin dependence	*E. coli*	1×10^{-10}
Streptomycin resistance (25 µg)	*Hemophilus (Bordetella) pertussis*	6×10^{-10}
Streptomycin resistance (1000 µg)	*H. pertussis*	1×10^{-10}

*From Braun, W.: *Bacterial Genetics,* 2nd edition, Philadelphia, Saunders, 1965. By permission of the publishers.

Factors Affecting Estimations of Mutation Rates

Some specific factors that induce errors in estimating mutation rates are as follows:

1. *Differential Growth Rates of Parent and Mutant*—Most methods for estimating mutation rates assume that the cultural conditions utilized favor neither the parent cells nor the mutant cells under study. This is not necessarily the case. Mutants frequently exhibit altered growth rates, particularly metabolic mutants, resulting in errors in measuring mutation rates.

2. *Lags*—This multitude of factors is responsible for the most serious errors in mutation rate estimations.

a. *Phenomic Lag*—if an organism suffers a mutation that causes it the loss of the formation of an enzyme, for example, residual amounts of the enzyme will still be in the cytoplasm of the cell even after one or more divisions. It is not until the enzyme molecules are diluted out among the daughter cells that the cells finally exhibit the mutant phenotype (Fig. 59), leading to an overestimation of the rate at which the mutation occurs.

b. *Phenotypic Lag*—rapidly growing bacteria frequently have as many as four nuclei per cell. If a mutation occurs in one of the nuclei just before cell division, and the mutant character is recessive, it may be three generations before the mutant phenotype is expressed (Fig. 60).

c. *Segregation Lag*—in the case of a dominant mutation in multinucleated cells, the phenotype may be expressed immediately, but no increase in the number of mutant cells will be observed until several generations later (Fig. 61). Both segregation and phenotypic lags will create underestimates of mutation rates.

3. *Residual Growth of Parents*—Additional mutations may occur in parent cells after selective conditions are imposed upon them. This will have a tendency to create overestimations of the mutation rate (Fig. 62).

4. *Back mutations*—As we will discuss in Chapter Five, a mutant cell can experience a subsequent mutation that has the effect of reversing the effects of the original mutation. Thus, a mutation may go undetected because it has been reversed before it is discovered. Underestimations would be the obvious results of back mutations.

5. *Inconstant Mutation Rates*—It is almost always assumed that the rate at which a given mutation occurs is constant throughout the duration of an experiment. This may not be the case. We will have occasion to refer to the existence of mutator genes, genetic regions that on experiencing mutations themselves will bring about an increase in other mutation rates.

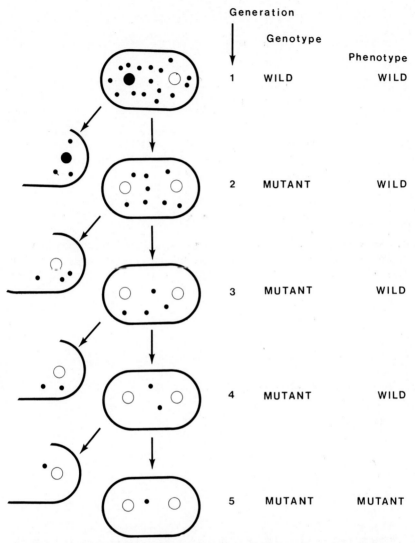

Figure 59. Example of the effect of phenomic lag on the determination of mutation rates. In generation 1, a bacterium experiences a mutation that blocks the formation of a particular enzyme (small circles). In generation 2, the mutant cell has divided and is now homozygous with respect to the mutant gene. However, it still retains the wild phenotype because of residual enzyme in the cytoplasm. This situation may prevail for two or more generations, until as in our case, in generation 5, the enzymes have been diluted out to the point that the cell now exhibits the mutant phenotype.

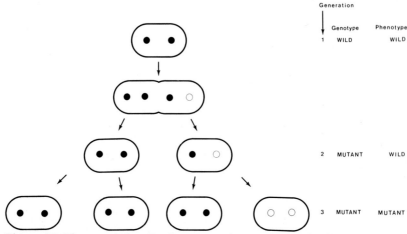

Figure 60. Phenotypic lag. During generation 1 a bacerial cell experiences a recessive mutation, but it is not expressed until generation 3 when the homozygous state is finally achieved.

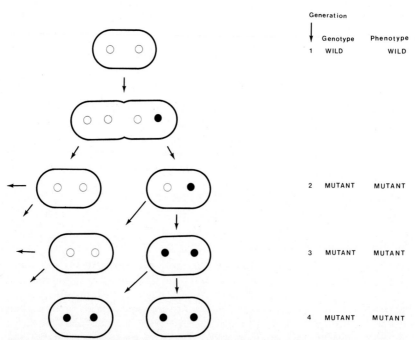

Figure 61. Segregation lag. A bacterial cell undergoes a mutation with a dominant expression in generation 1, resulting in an immediate mutant phenotype. The number of cells expressing the mutant phenotype does not change until generation 4, however, causing an error in the determination of exactly when the mutation actually occurred.

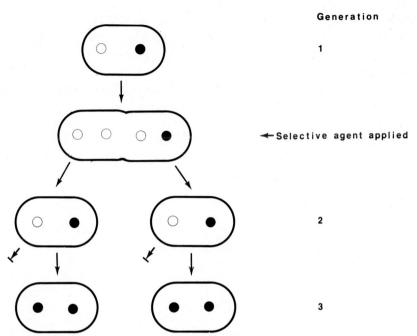

Figure 62. Residual growth of parent cells. During generation 1, a cell experiences a mutation that affords it resistance toward some selective agent, such as an antibiotic. The selective agent is applied near the end of generation 1, but one of the parental cells escapes the effects of the selective agent long enough to experience a mutation in generation 2 that affords it resistance. Since it is presumed that all sensitive parental cells were eliminated by the antibiotic in generation 1, the second resistant clone is thought to have originated from a mutation prior to the application of the selective agent. An overestimation of the rate at which the mutation occurs results.

There is always a reasonable probability that an organism will experience a mutator mutation during an experiment.

6. *Mutation Repair and Fixation*—Some capability to repair damage inflicted on their DNA has been demonstrated in living cells at all phylogenic levels. Bacteria possess several repair systems, the details of which will be described in a later section. Repair capacity is under genetic control, but it also is under some influence by the physiological state of the cell. That is to say, under one set of physiological conditions, a premutational

lesion, such as a thymine-thymine dimer, may be repaired before a base transition or other alteration is induced. Under other conditions, DNA replication may occur before repair mechanisms have had the opportunity to act and a mutation may be fixed in the DNA sequence. Evelyn Witkin was the first to show such a *mutation fixation* phenomenon, in which the nature of the medium into which cells were transferred had a great bearing on the number of UV-induced mutations that ultimately appeared in the culture (Fig. 63). This source of error in determining mutation rates is one of the strongest

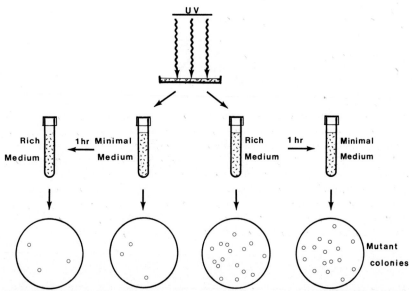

Figure 63. Mutation fixation. If a suspension of bacteria is irradiated with ultraviolet light, the number of mutant colonies that form on subsequent agar plates depends upon the nature of the medium in which the irradiated cells are incubated immediately following irradition. If the cells are transferred to a rich medium, a maximum number of mutant colonies form. Transfer to a minimal medium results in a reduced number of mutant colonies. The effect occurs within one hour of irradiation, for if the cells are subcultured from minimal to rich, or from rich to minimal medium after an hour's incubation, the outcome is not changed with respect to the first medium in which the organisms are incubated.

arguments for carrying out such experiments under carefully standardized conditions.

Subsequent studies of mutation fixation have shown that the key to it involves protein synthesis. Irradiated cells that have had protein synthesis briefly inhibited exhibit a smaller number of mutations than cells that are grown in a medium encouraging rapid protein synthesis. The phenomenon in which mutation frequencies are lowered as a result of temporary protein synthesis inhibition has been named *mutation frequency decline* (MFD).

Students frequently ask why there are so many methods for estimating mutation rates, each apparently resulting in a different value. There is no one ideal method for the estimation of mutation rates. All of the methods that have been proposed are based on one or more assumptions and approximations. Estimations based on formulas (5) and (8) tend to be low, whereas formula (7) results in higher estimates. For an excellent discussion of these and other methods of estimating mutation rates, the reader is urged to consult Newcombe's 1948 paper.

It is clear that knowledge of mutation rates is of limited value. Their most useful application is in comparing the efficiencies of various mutagens, or comparing mutation rates among various strains of the same species of organism.

A recent example of such an application of mutation rate estimation is in the case of bacteriophage Mu-1. This bacteriophage was thought to induce mutations in certain *E. coli* host strains. The phage does not lyse its host bacteria, but instead, its DNA on entering the bacterial cell becomes integrated into the DNA of the host, establishing a relationship known as lysogeny (Chapter Seven). Genes adjacent to the site of phage DNA integration, including one responsible for T1 phage resistance, appeared to increase their rates of mutation. Geneticist A. L. Taylor measured the occurrence of T1 resistance in host bacteria in the presence and absence of integrated Mu-1 DNA, and by applying method (8), was able to show that the presence of Mu-1 DNA was indeed mutagenic (Table XIV).

TABLE XIV
EVIDENCE OF THE MUTAGENIC EFFECT OF Mu-1 PROPHAGE*

Uninfected control culture:

M_1 = 14 T1 phage-resistant colonies
M_2 = 74 T1 phage-resistant colonies
N_1 = 0.65x10^8
N_2 = 23.2x10^8
Time = 110 minutes
Calculated mutation rate[†] = 1.8×10^{-8}

Mu-1-infected culture:

M_1 = 21 T1 phage-resistant colonies
M_2 = 149 T1 phage-resistant colonies
N_1 = 5.4x10^6
N_2 = 83.3x10^6
Time = 120 minutes
Calculated mutation rate[†] = 1.14×10^{-6}

*From Taylor, A. L.: *Proc Natl Acad Sci USA, 50:*1043, 1963.
[†]By the method of Newcombe (see text).

Mutator Genes

In the previous section we related that some, if not most, spontaneous mutations probably are caused by errors in DNA replication or repair. Occasionally, strains of bacteria, bacteriophage or other microorganisms exhibit sudden, overall increases in observed spontaneous mutation rates by factors of less than ten to many thousand. The source of the increases can be traced to mutations occurring in specific regions of the genome, regions designated as *mutator genes*. A mutator gene may be involved in the formation of DNA polymerases or other enzymes responsible for DNA replication. A mutation in such a gene could lead to a greater number of errors in replication, errors that eventually may lead to mutations. Alternatively, the mutator gene may be one that oversees the biosynthesis of certain DNA precursors such as the purines and pyrimidines. A mutation here could result in errors that are qualitative (alterations in base structure) or quantitative (alteration in number of molecules produced), in either case creating situations that could increase the chances for mutations. As we will see in the next section, DNA repair processes play an important role in the survival of organisms. Any mutational damage sustained in a genetic region involved in DNA repair also could lead to an increase in spontaneous mutation rates.

DNA REPAIR

Photoreactivation

In 1949, A. Kelner reported that if he exposed suspensions of bacteria that had just been exposed to lethal doses of UV light to strong white light, the lethal effects of the UV irradiation could be reversed by as much as 90 percent. This phenomenon, now known as *photoreactivation,* is catalyzed by a light-activated enzyme that reduces pyrimidine dimers to the monomeric state and thereby restores the DNA to its normal configuration. Photoreactivation apparently is a widespread capability among bacteria, fungi, protozoa, algae, and some cultured animal cells. The action of photoreactivation is specific in that it repairs only damage induced by UV light, and only pyrimidine dimers. The enzyme responsible for this activity has been purified and is active towards irradiated DNA *in vitro.*

Subsequently it has been discovered that bacteria possess at least three general systems for the repair of various types of damage to DNA. In addition to photoreactivation, there also is *dark* or *excision repair,* and *postreplication repair.*

Excision Repair

Another DNA repair system possessed by bacteria operates independently of a source of white light. Known as *dark* or *excision repair,* this system appears to involve at least three enzymes. When a lesion, such as a thymine-thymine dimer, is repaired, the first step is the removal of the offending pyrimidines by an enzyme that breaks the single DNA strand one or two nucleotides on either side of the dimer (Fig. 64). Additional nucleotides may subsequently be removed to enlarge the gap even more. By using the opposite strand as a template, a second enzyme now patches the gap with complementary nucleotides, and a third enzyme seals the single-strand break between the new nucleotide sequence and the

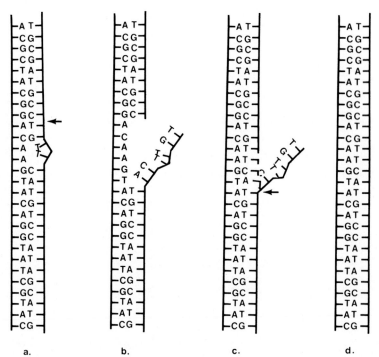

a.   b.   c.   d.

Figure 64. Excision, or dark repair of DNA. An endonuclease removes the damaged section of DNA (a.), after which a polymerase restores the original nucleotide sequence by utilizing the opposite strand as a template (c.). The damaged section is finally removed entirely, and a third enzyme then completes the repair process (d.) by forming the last phosphodiester bond.

original strand. In addition to differing from photoreactivation in its light independence, excision repair also shows repair capability over various types of damage induced by nitrogen mustard, NTG, mitomycin C, and other agents as well as that by UV light. Excision repair is inhibited by caffeine.

Postreplication Repair

A third repair system, known as *postreplication repair,* appears not to repair immediately the primary lesion induced by UV light but to repair the damage through a mechanism of recombination and repair following replication of the DNA

(Fig. 65). While some postreplication repair steps resemble the last steps of excision repair, it has been shown that each system is under the control of distinctly different genes. Excision repair is under the control of the genetic locus known as *uvr,* whereas postreplication repair appears to be associated with one of several genes involved with the ability to support recombination. The controlling gene for postreplication repair is known as *rec*A. Furthermore, postreplication repair seems to be based on a recombination mechanism rather than the patch and cut process of excision repair. It appears that errors made during postreplication repair account for most ultraviolet light-induced mutations.

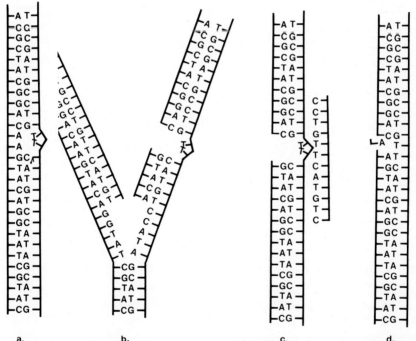

a.   b.   c.   d.

Figure 65. Post replication repair. At times, a thymine dimer resulting from ultraviolet irradiation (a.) will not be repaired before DNA replication occurs (b.). As a result, a gap is left on the strand opposite the lesion. A strand homologous to the damaged strand then undergoes recombination with the latter (c.), restoring the original sequence. The gap left opposite the dimer is now filled by other repair enzymes.

Chapter Five

MUTATIONS II

MUTANT PHENOTYPES: MOLECULAR LEVEL

W E HAVE SPOKEN OF mutations as changes in the genotypic information of a cell, but we have not detailed the effects of such changes on the phenotype of the cell. The fact of the matter is, a mutation may result in anything from no effect to a complete loss of a function, or death, for the cell. If one considers only point mutations in a structural gene for the moment, those that involve a change of a single base pair, the outcome is of course the creation of a new triplet codon. The consequences of such an alteration are summarized in Table XV. Because of the degeneracy of the genetic code, a cell may be able to absorb certain base changes with no effect; the same amino acid is coded by the mutant codon.

In cases where the mutant codon denotes a different amino acid, the substitution may be permissive in that for any of a number of reasons the change does not significantly alter the function of the involved polypeptide. In some instances the new amino acid may not be a completely good substitute, frequently resulting in what is known as a *leaky mutant*. A leaky mutant is one that has suffered a reduction in the activity of a given protein but not its complete elimination. In fact, the same number of protein molecules may be formed. The term was originally coined to denote an enzyme being only partially turned off (analogous to a "leaky" faucet), but this is misleading in that the enzyme's activity frequently is turned off qualitatively, not quantitatively.

The alteration in the phenotypic expression of a protein may

TABLE XV
CONSEQUENCES OF POINT MUTATIONS IN BACTERIA AND BACTERIOPHAGES

What happens when a single base-pair is changed within a gene which is responsible for the formation of a given polypeptide?

The resulting change can:

1. Code for the same amino acid
   (no effect on polypeptide activity or function)
2. Code for a different amino acid (= missense mutation)
   a. *no effect*
      —new amino acid is similar enough to original that it can substitute for it, e.g.:

glu	lys	gly	val
↓↑	↓↑	↓↑	↓↑
asp	arg	ala	leu

      —new amino acid not in critical location on polypeptide chain
   b. *reduction in activity*
      —new amino acid not completely good substitute (= leaky mutant)
   c. *increase in activity*
      —new amino acid improves activity of polypeptide through more favorable folding or other structural alterations
   d. *complete loss of activity or function*
      —substitute amino acid is in critical location, involving active site, folding, disulfide bridges, etc.
3. Code for no amino acid (= nonsense mutation)
   —result of amber-type mutation, leading to a UAA, UAG, or UGA codon on the mRNA and premature termination of chain elongation
   CRM⁺ = enough polypeptide chain produced to be detected serologically
   CRM⁻ = polypeptide chain too short to be detected serologically

be conditional in that the protein expresses the mutant phenotype only under certain conditions. *Temperature sensitive (ts)* mutations are an excellent example of such conditional mutations. The structural alteration of the protein as a result of an amino acid substitution may have no effect at normal temperatures, but may lose its functioning partially or completely at a slightly higher temperature. Some *ts* mutations are lethal, an example of which is seen in the case of certain strains of *E. coli,* which can grow at 37° C but not at 42° C.

A mutation may lead, although rarely, to an increase in the activity of a given protein. In the case of an enzyme, it may gain a wider substrate range. The implications of this kind of mutation to evolution are far-reaching.

Finally, a given amino acid substitution may be a complete failure, where the affected protein has lost all activity.

In a special case, the mutant codon may not code for any

amino acid, but may be one of the three punctuation, or nonsense codons UAA, UAG, or UGA. The appearance of one of these mutations in the middle of a structural gene will result in the premature termination of the elongation of the polypeptide chain. This has been shown by analyzing the resulting polypeptides made by various mutants suffering nonsense mutations in the gene for that polypeptide (Fig. 66). By genetic means, it was possible to locate the exact position of each nonsense mutation on the gene. The length of the polypeptide chain produced by each mutant was demonstrated to be proportional to the distance between the mutation site and the region of transcription initiation on the gene. Cells that have suffered a nonsense mutation may still produce enough of the polypeptide chain length for it to be detected serologically with specific antiserum. Such mutants are known as being CRM$^+$, or Cross-Reacting Material positive. Some mutants may have experienced the mutation so close to the initiation end of the structural gene that the short fragment of polypeptide that is formed is too small to be detected serologically. These mutants are designated as CRM$^-$.

Regulatory genes also may experience mutations. Referring back to our discussion of the operon in Chapter Three, mutations have been detected in all of the regulatory genes involved there. For example, the *i* gene may suffer a mutation that eliminates its ability to make the repressor. The operon finds itself permanently derepressed, that is, constitutive. Another mutation mapping in the *i* gene is called i^s, for superrepressed. In this situation, the mutation apparently has altered the structure of the repressor such that it no longer reacts with inducer. The result is that the operon is permanently repressed and makes no protein either in the presence or absence of inducer. The operator also is subject to mutations. One results in the *o* region being unable to react with repressor, resulting again in a permanent state of derepression. The designation for this mutation is o^c, c denoting a constitutive mutant. Another mutation located in the *o* region is the o^0 (o-zero, or operator-negative) mutation. This lesion appears to be a nonsense mutation that prevents the

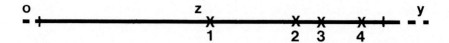

Mutant	Type	Molecular Weight (Observed)	Molecular Weight (Calculated)
1	Amber	60,000	67,000
2	Amber	91,000	101,000
3	Amber	103,000	111,000
4	Ochre	122,000	127,000

Figure 66. Four mutants of *E. coli*, each suffering a nonsense mutation in the β-galactosidase structural gene (z) of the lactose operon, were examined for the length of the incomplete protein produced. The observed molecular weights of the protein fragments were estimated by ultracentrifugation in a sucrose gradient. These values were compared with molecular weights calculated on the basis of the position of the mutation on the gene and on the molecular weight of the complete protein, 135,000. These results confirm the notion that transcription begins at the end of the gene adjacent to the operator region *(o)* and that the effect of a nonsense mutation is the premature termination of polypeptide chain elongation. (From Fowler, A. V., and Zabin, I: *Science, 154:*1027, 1966. Copyright 1966 by the American Association for the Advancement of Science)

operator gene from activating the promoter gene and initiating transcription.

From this discussion, it is clear that mutations may bring about a number of effects. As a rule the only mutations we learn of are those that are expressed phenotypically. This brings us to the point where it is necessary to sharpen our concept of the definition of phenotype. We have defined phenotype as the outward expression of the genotype. It should be made clear that outwardness includes the demonstration of a change in a single amino acid in an otherwise normally functioning chain of 400 amino acids, a demonstration requiring great allocations of equipment and manpower, as well as the loss of pigment production in a culture of *Serratia,* observable by any freshman bacteriology student.

Forward and Back Mutations

We have described mutations as changes without designating the direction of the change. That is to say, if an organism gains or loses a function as a result of a mutation, was it normal for the organism to have the function in the first place? The "normal" state in genetics terminology is designated as the *wild* state. In microorganisms, the wild state is generally that phenotype which is expressed by a specified strain, usually a type species deposited in a recognized culture collection, such as:

American Type Culture Collection
Rockville, Maryland

Northern Regional Research Laboratory
U. S. Department of Agriculture
Peoria, Illinois

Thus, a mutation that departs from the wild phenotype is referred to as a *forward mutation.* A *back mutation,* or *reversion,* is a mutation that restores the wild phenotype to the organism.

Much can be learned about the origin of a given mutation by studying the nature of and the conditions that bring about its reversion. As a general rule, those mutations that exhibit high rates of reversion are the point mutations described earlier: transitions and transversions. Thus, through a spontaneous mutation an A-T pair may undergo a transition to a G-C pair.

There is virtually nothing to prevent the G-C pair from spontaneously reverting back to an A-T pair in some subsequent generation. This would be an example of a *true back mutation.* Less likely to revert are the small deletions or additions that produce mutations of the frame-shift variety. Large deletions obviously would be the least likely to revert. By knowing the specific action of a mutagen such as nitrous acid, for example, that is known to bring about the reversion of a given spontaneous mutation, it may be possible to deduce the nature of the original mutation.

Suppressors

A detailed study of the mechanisms of reverse mutations is highly complex, for a reversion may not necessarily involve the particular base-pair responsible for the original mutation, nor even involve the same gene. A back mutation that involves a site other than that of the original mutation is referred to as a *suppressor mutation.*

Suppressor mutations may operate in either a direct or an indirect manner. A *direct suppressor* would be one that directly restores the function of the mutated gene. Direct suppressors may be *intragenic,* in which case they occur in the same affected gene, or they may be *intergenic*—in quite another gene. The best example of a direct, intragenic suppressor is the instance of the acridine-induced deletions utilized by Crick to determine the size of the reading frame in his genetic code studies. The third base deletion or addition, which restored the reading frame, would be considered a direct, intragenic suppressor mutation.

1. *Direct, Intergenic Suppressors*—Direct, intergenic suppressors can revert missense, nonsense and frame shift mutations. The source of this class of reversions generally has been shown to be mutations that affect components of the polypeptide synthesizing machinery of the cell, particularly tRNA. Basically the suppressor mutations bring about alterations in the structures of these components such that a missense or nonsense triplet is misread as another amino acid. Missense

suppression is brought about by the chance substitution of an amino acid more fit in that location than the mutational amino acid, but usually less fit than the original (Table XVI). This latter point is important in that a suppressor frequently does not restore the affected gene to its full, wild state of efficiency, but may result in a *pseudowild* reversion.

TABLE XVI
PROPERTIES OF SOME SUPPRESSOR MUTATIONS IN *ESCHERICHIA COLI**

Symbol	Location on Chromosome[†]	Mutant codon suppressed	Amino acid inserted	Efficiency (%)
Sup D	38	UAG	serine	60
Sup E	15	UAG	glutamine	30
Sup F	26	UAG	tyrosine	50
Sup G	16	UAA	lysine	6
Sup O	26	UAA, UAG	tyrosine	12 (UAA) 16 (UAG)
Sup U	75	UAG	glutamine	76

*From Smith, J. D.: *Annu Rev Genet,* 6:235, 1972. Reproduced with permission of Annual Reviews, Inc.
[†]See Figure 100.

More detailed information has been gathered regarding nonsense suppressors than for missense suppressors. For example, coliphage T4 has been known to suffer a nonsense mutation in which the untranslatable amber codon UAG is created in the middle of a mRNA coding for a particular polypeptide. Loss of this polypeptide prevents the phage from lysing certain strains of *E. coli.* In spite of the mutation, these amber phage mutants can infect and lyse other strains of *E. coli* that, on further inspection, have been found to have experienced suppressor mutations. One result of a suppressor mutation is the structural alteration of the anticodon on a host tRNA. For instance, the normal anticodon on tyrosine tRNA is $G^{OMe}UA$, which complements the tyrosine codon UAC. * Bacteria that carry the *sup*F suppressor mutation possess low levels of an altered tyrosine tRNA that has as its anticodon CUA. CUA would be an anticodon for the amber codon UAG.

*Nucleotide sequences are always read from the 5′ end; in base-pairing, this codon-anticodon pair would appear as $G^{OMe}UA$.
                                                                $\overline{\text{C \quad AU}}$

Therefore, when an amber T4 mutant injects its DNA into the *sup*F carrying host, the amber codon on the phage mRNA occasionally will be read as tyrosine rather than as a termination signal, and some synthesis of the polypeptide encoded in the mutant mRNA will therefore be completed. Whether tyrosine was the original amino acid at the mutation site remains to be seen, but at least it is a good enough substitute to allow the phage to complete its infectious cycle in the host. Note that it is the presence of the suppressor in *the host* that suppresses the mutation of *the phage*. Bacteria can suppress mutations in their own genome as well, however. Table XVI lists examples of several nonsense suppressors. The listed efficiencies are based on estimates made of the actual amount of protein synthesized by the suppressed mutant gene, whether the protein proves to be functional or not. That is to say, protein quantities are measured serologically by a quantitative immunochemical reaction that is independent of the intrinsic activity of the protein.

Since suppressor mutations are specific in that an amber suppressor suppresses only amber mutations, the use of *sup* hosts has been a valuable tool for screening large numbers of phage particles for the presence of specific nonsense mutations.

Altered tRNA's also have been shown to be responsible for the suppression of frame-shift mutations, as well. The mechanism of such suppression is not at all clear.

2. *Indirect, Intergenic Suppressors*—Indirect, intergenic suppressors frequently operate by opening up alternate metabolite pathways to bypass the paths blocked by mutation.

As an example, *Neurospora* synthesizes its aromatic amino acids plus paraaminobenzoic acid from shikimic acid (SA). Shikimic acid is formed from dehydroshikimic acid (DHS), the step being catalyzed by the enzyme dehydroshikimic acid reductase (Fig. 67). Loss of the ability to form this enzyme through mutation thrusts the organism into a dependency on aromatic amino acids in the medium. Contrary to other similar situations however, dehydroshikimic acid does not accumulate in the cells in any great quantity but is converted to protocatechuic acid by a second enzyme, DHS-dehydrase.

Suppressor mutations have been detected in *Neurospora* that restore its aromatic amino acid independence, and that map in the gene responsible for the formation of DHS dehydrase. This fungus is known to possess a second, inducible enzyme that can convert dehydroshikimic acid to shikimic acid. Presumably when the organism loses its ability to convert dehydroshikimic acid to protocatechuic acid, the former compound does accumulate in sufficient quantities to bring about its conversion to shikimic acid via the inducible pathway.

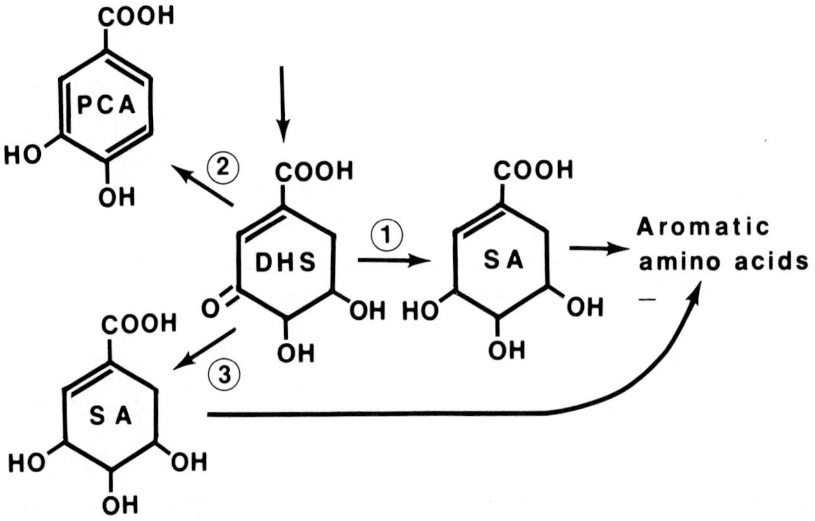

Figure 67. Example of an indirect, intergenic suppressor mutation. *Neurospora* has been shown to produce its aromatic amino acids from shikimic acid (SA), which is derived from dehydroshikimic acid (DHS) through a reaction catalyzed by the constitutive enzyme dehydroshikimic acid reductase (1). The fungus may lose the activity of enzyme 1 through a mutation, but the accumulation of DHS is prevented by the activation of inducible enzyme 2 to form protocatechuic acid (PCA). PCA cannot act as a precursor for amino acids, however. A second mutation can occur that blocks the activity of enzyme 2, in which instance the accumulation of DHS initiates a second inducible enzyme, 3. Enzyme 3 forms shikimic acid, which can then be converted to the needed amino acids. The mutation that involves enzyme 2 would be referred to as a suppressor mutation, for it has the effect of reversing the first mutation that intervened in enzyme 1 activity and of restoring the organism's independence for aromatic amino acids. (From Case, M., Giles, N. H. and Doy, C. H., *Genetics, 71*:337, 1972).

MUTANT PHENOTYPES: ORGANISMIC LEVEL

For decades, various mutations have been observed in bacteria and other microorganisms. In *E. coli,* the locations of hundreds of mutations have been mapped on the organism's chromosome. Nearly 95 percent of these mutation sites can be grouped into five more or less arbitrary classes according to their phenotypic expression. In the following sections we will deal with at least one type of bacterial mutant from each of the five classes, including methods for their isolation. The classes and the examples to be covered are shown in Table XVII.

TABLE XVII
CLASSIFICATION OF BACTERIAL MUTATIONS

Class	*Examples discussed in this chapter*
I. Sensitivity towards selective agents	Antibiotic resistance Ultraviolet radiation resistance
II. Anabolic	Auxotrophy
III. Catabolic	Fermentation
IV. Relationships with bacteriophages	Bacteriophage resistance
V. Structures and functions	Flagellar structure Recombination deficient Aberrant division (minicells) Spore formation

Before specific mutants are discussed, a presentation of some general principles regarding their isolation in the laboratory is in order. Knowledge of techniques for the isolation of various microbial mutants is one of the most important lessons a student can come away with. Rather than include minute details, which can be found in at least two excellent manuals (Clowes and Hayes, and Miller) and other references to be cited, the author has chosen to present broad outlines of each technique to enable the reader to grasp the basic principles of mutant isolation.

When presented with the task of singling out organisms possessing a specific mutant phenotype from a normal, growing culture of bacteria, one must realize that the mutants frequently will be outnumbered by the wild type by ratios of 10^6 or more. This ratio is of course based on the rate of occurrence of the specific spontaneous mutation. The ratio can be improved considerably by the application of one or more of the mutagens described in Chapter Four. Additional enrichment of the mutant cells occasionally can be brought about by the use of methods such as the penicillin technique and others described in this chapter.

However, increasing the relative numbers of the mutant cells is less than half the battle, for one must still find ways of identifying a mutant cell against a background os perhaps 10^4 wild type cells and isolating it in pure culture. In practical terms this could mean the picking and testing of 10^4 colonies before the mutant is identified.

If the nature of the sought-after phenotype affords the mutant cells a strong selective advantage over the wild type, as in the case of bacteriophage resistance, elimination of the wild type and isolation of the mutants is a relatively easy affair. Situations in which the selective pressure is somewhat less powerful, such as in antibiotic and radiation resistance, still can be resolved with special techniques. Where perseverance, ingenuity, and sheer luck play a greater part is in the isolation of mutants that experience neutral or negative selective pressure relative to the wild type. Certain nutritional mutants fall into this classification. The Lederberg replica plating method (Chapter Four) offers considerable aid in this case. Its application to the isolation of auxotrophs and other types of mutants is described in later sections. However, in some situations replica plating or other stepsaving techniques are not applicable and the geneticist has no choice but to screen 10^6 or more colonies on appropriate differential media for signs of the mutant phenotype.

Antibiotic Resistance

The development of resistance towards chemotherapeutic drugs in microorganisms was recognized almost from the very beginning of the use of such compounds. The same year (1907) Paul Ehrlich reported on the trypanocidal action p-rosaniline, his colleagues discovered that the protozoa could develop resistance towards the dye. The first report of drug resistance in bacteria appears to be in 1912, when optochin-resistant pneumococci were observed.

It is widely known that thousands of strains of antibiotic-resistant bacteria have appeared in clincial isolations in more recent times. Resistance towards antibiotics among bacteria is not an absolute response. Sensitivity and resistance are relative terms, the confines of which are defined by the specific strain of organism and the specific antibiotic under consideration. A Gram-positive bacterium known to be sensitive to penicillin may be killed by a concentration of 1 to 2 units/ml, whereas a Gram-negative bacterium still may be considered sensitive even if it takes 1000 units/ml to kill it. The point is that these concentrations define the minimal inhibitory concentrations for the wild strains of the bacteria cited. However, these strains through various means may acquire the ability to survive and grow in the presence of antibiotics at concentrations significantly higher than the minimal inhibitory concentrations. One general means by which a bacterium may gain antibiotic resistance is through mutation.

Antibiotic resistance frequently manifests itself in two distinct patterns: multistep and single-step (Fig. 68). Resistance towards penicillin follows a multistep pattern in that an organism may acquire resistance to greater and greater levels of the antibiotic through a series of specific mutations. Thus, an initial mutation may afford the bacterium and its progeny resistance towards 4 units/ml. Several generations later a subsequent mutation may increase the organism's tolerance to 8 u/ml, and each additional mutation will add its expression to the phenotype of the bacterium in the form of increased resistance.

In the case of streptomycin resistance, a single-step pattern emerges. Here, once an organism experiences its first mutation to streptomycin resistance, subsequent mutations in genes responsible for streptomycin resistance have no effect. The presence of such additional mutations can be demonstrated by genetic means, even though their phenotypic expression is not manifested in the presence of the first, original mutation.

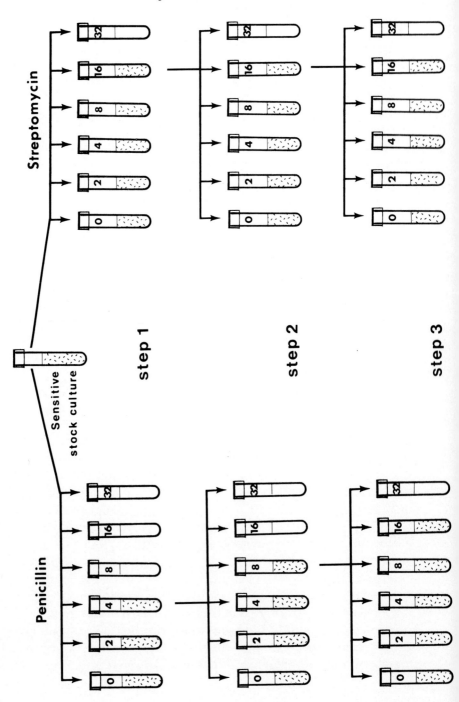

step 4

Figure 68. Multistep and single-step acquisition of antibiotic resistance. In the case of penicillin, a sensitive bacterial strain may acquire resistance to 4 units/ml through a mutation in the first subculture. An inoculum taken from this resistant culture may yield organisms that experienced a second mutation that afforded them resistance to 8 units/ml, and in each subsequent subculture, additional mutations affording the bacteria higher levels of resistance may occur. This is referred to as a multistep resistance pattern. In the example of streptomycin, the stock culture may produce mutant cells that are resistant to 16 μg/ml. On subsequent subculturing, mutants exhibiting higher levels to tolerance cannot be isolated. This behavior reflects a single-step resistance pattern.

Bacteria appear to possess several multiple-factor genes each responsible for resistance towards specific antibiotics. In the case of resistance towards penicillin and most antibiotics, as each gene experiences a mutation, its phenotype is added to the phenotypes of prior mutated resistance genes, resulting in an accumulation of resistance levels.

The reason for the expression of subsequent streptomycin resistance mutations blockage is unknown. Other drugs exhibiting the single-step pattern are isoniazide, erythromycin, and paraaminosalicylic acid.

Two additional responses towards antibiotics exhibited by microorganisms are *cross-resistance* and *collateral sensitivity* (Table

TABLE XVIII

CROSS RESISTANCE AND COLLATERAL SENSITIVITY
IN *ESCHERICHIA COLI*

Source –	bacteria					streptomyces						fungi	
	polypeptides												
	Bacitracin	Polymyxin B	Circulin	Viomycin	Streptothricin	Neomycin	Catenulin	Streptomycin	Netropsin	Chlortetracycline	Terramycin	Chloromycetin	Penicillin
E. coli strain													
wild	1	1	1	1	1	1	1	1	1	1	1	1	1
A	1	1	1	1	1	1	1	1	1	1.5	1.5	1.5	1.5
B	20	1/20	1/100	1/1.5	1/1.5	1.1	1	1	1	1	1/2	1	1.5
C	1	20	10	1	4	4	3	10	2	1	1	1	1/1.5
G	1	1/2	1/10	2	15	70	15	5	1	1	1	1/1.5	1/2
H	1	1.1	1	1	1	1	1.1	1	100	30	10	10	60

Numbers refer to antibiotic resistance or sensitivity relative to the wild strain.

1 = same as wild strain; greater than 1 indicates resistance and approximate magnitude; less than 1 (fractions) indicates increased sensitivity relative to the wild type.

(From: Szybalski, W., and Bryson, V.: *J Bacteriol, 64*:489, 1952)

XVIII). Cross-resistance refers to the observation frequently made in which the acquisition of resistance towards one antibiotic is accompanied by resistance towards one or more additional antibiotics. Suggested bases for cross-resistance are:

1. *Chemical Similarities*—Several drugs are similar enough chemically that one would expect cross-resistance between them. For example, Terramycin® and Aureomycin® are structurally very much alike (Fig. 69), and in fact frequently exhibit cross-resistance. But cross-resistance is also seen in the case of chloramphenicol and penicillin, two drugs whose mode of action and structure are quite disimilar (Fig. 70). The reason for this anomaly is unknown.

2. *Alternate Metabolic Pathways*—An organism may require two or more parallel metabolic pathways to synthesize an essential growth factor (Fig. 71), with a different antibiotic affecting each pathway. Through a mutation the organism may

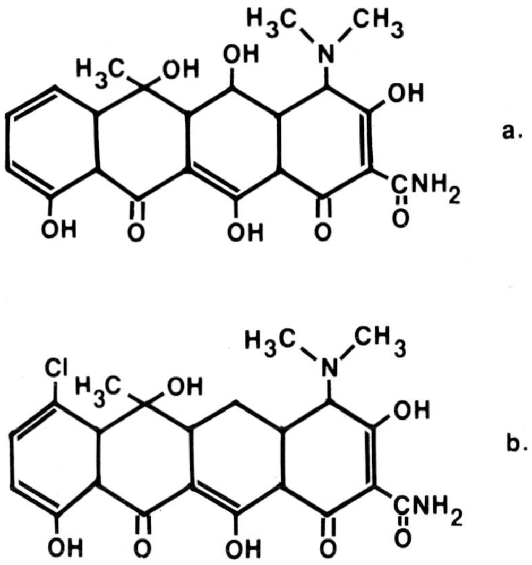

Figure 69. Structures of Terramycin® (a.) and Aureomycin® (b.). (Terramycin and Aureomycin are registered trademarks of Chas. Pfizer and Co., Inc., and Lederle Laboratories Division, American Cyanamid Co., respectively).

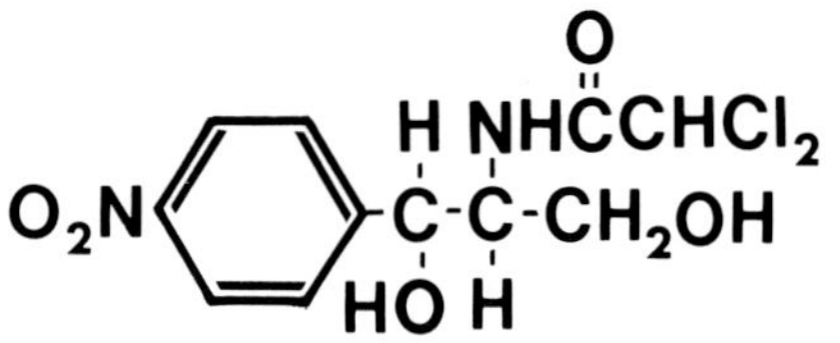

Figure 70. Structures of chloramphenicol (a.) and penicillin (b.). Dashed line in penicillin structure indicates bond that is broken by penicillinase (β-lactamase).

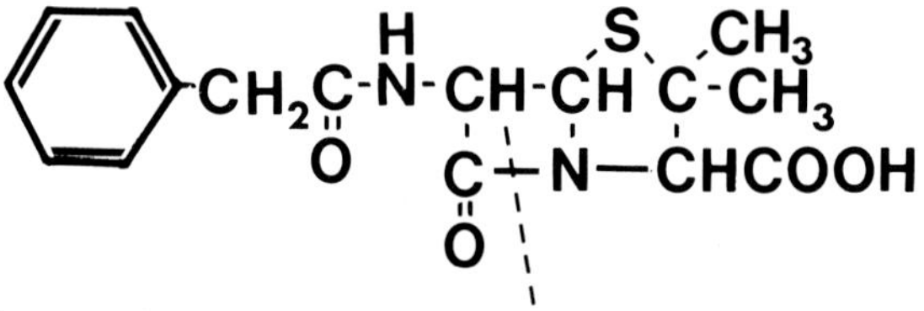

gain an alternate pathway that bypasses the one that is blocked and thus acquire resistance to all drugs involved.

3. *Nonspecific Changes*—Changes in cell membrane permeability that offer drug resistance might be a basis for cross-resistance in that the permeability block may be against all compounds of a general structural shape or charge distribution.

In collateral sensitivity, an organism may acquire resistance towards one antibiotic, and simultaneously lose resistance towards another. An example of collateral sensitivity is in the case of Strain B (Table XVIII). This strain gained resistance towards bacitracin, but at the same time lost resistance towards seven other antibiotics, including penicillin. The mechanism of this action is unknown.

Mechanisms of Antibiotic Resistance

A detailed discussion of the mechanisms of antibiotic resistance is beyond the scope of this book, but a brief description of them is in order:

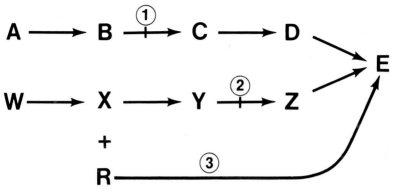

Figure 71. Alternate metabolic pathways. Reactions 1 and 2, each affected by a different antibiotic, are needed to form the essential metabolite, E. A mutation may occur that opens up a new pathway, 3, so that the organism no longer is dependent on reactions 1 and 2. The organism thus becomes resistant to both antibiotics *simultaneously*.

Antibiotics act on microorganisms by inhibiting critical functions, generally the synthesis of essential precursors or macromolecules, or the maintenance and function of structural entities, such as the ribosomes or the cell wall. In the acquisition of resistance towards a specific antibiotic, the organisms appear to regain the blocked functions in spite of the presence of the antibiotic. Some basic mechanisms of antibiotic resistance can be outlined as follows:

1. *Detoxification*—Bacteria may achieve the ability to inactivate the antibiotic through derivative formation. For instance, resistance towards certain antibiotics has been shown to be due to the formation of one or more enzymes, the best known of which is penicillinase. The activity of this enzyme is the cleavage of the lactam ring in the penicillin and *Cephalosporin* molecules, destroying the antibacterial action (Fig. 70). The penicillinase of Gram-positive bacteria generally is inducible and is released into the medium as an extracellular enzyme. Gram-negative bacteria produce a constitutive penicillinase that is cell-bound. Both enzymes, however, are identical in their effect on penicillin. Most drug-resistant bacteria isolated from clinical infections appear to have gained their refractoriness through

the mechanism of drug detoxification, whereas most bacteria from laboratory cultures that owe their resistance to the occurrence of spontaneous or induced mutations have acquired resistance through one of the following two mechanisms.

2. *Change in Target Molecule*—A mutation may induce a slight structural change in the molecule acted upon by the antibiotic. The change is not severe enough to alter significantly the molecule's function, but it is sufficient to reduce its affinity for the antibiotic. This happens in the case of streptomycin resistance. This antibiotic affects bacteria by reacting with the ribosomes and thereby blocking protein synthesis. The organism may experience a mutation that alters the structure of the ribosome such that it no longer complexes with streptomycin and thereby gains resistance towards it.

3. *Change in Permeability*—An antibiotic must be taken into the bacterial cell to act. Bacterial strains are known to have achieved resistance towards tetracycline because of a mutation-induced defect in cell membrane permeability that prevents its uptake.

4. *Increased Competition*—Many antibiotics and chemotherapeutic drugs are analogs of essential metabolites or cofactors and act by interfering with the normal functioning of biochemical pathways through competitive inhibition. Classic examples of such drugs are sulfanilamide and related compounds. The sulfa drugs resemble the vitamin paraaminobenzoic acid (PABA), an important cofactor in the synthesis of folic acid (Fig. 72). The resemblance is near enough that sulfanilamide successfully competes with PABA and thereby inhibits folic acid formation. Some sulfanilamide-resistant mutants of *Staphylococcus* isolated from clinical sources have been shown to have acquired the ability to produce some twenty times more PABA than the sensitive strains. The excess PABA affords the cells a more favorable competitive advantage over the drug, and thus, resistance towards it.

The genetic information that affords organisms the ability to execute one or more of the foregoing mechanisms can be found on the organism's chromosome, or on cytoplasmic, extrachromosomal bits of genetic material known as R (resis-

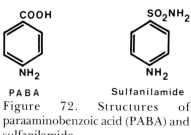

PABA　　　　**Sulfanilamide**

Figure 72. Structures of paraaminobenzoic acid (PABA) and sulfanilamide.

tance) factors. These virus-like particles contain genetic information for resistance towards a number of antimicrobial agents, including many of the antibiotics, UV light, and the inorganic ions of mercury and cadmium. R factors are self-replicating, can be transmitted from cell to cell by various mechanisms, and constitute a growing problem in clinical infections. They will be considered in greater detail in Chapter Seven.

Antibiotic Dependence

In a paradoxical case, a strain of *E. coli* has been found that is not only resistant to streptomycin, it is dependent upon it. That is to say, the organism will not grow unless streptomycin is added to the medium. It appears that a mutation-induced structural change in the 30s ribosome particle is so severe that it has rendered the particles inoperative. However, particles complexed with the antibiotic appear to have their activity restored, enabling the cells to resume protein synthesis. Bacteria also have been shown to develop dependence on other antibiotics that have as their target the 30s or the 50s ribosomal particles.

Isolation

Resistance towards antibiotics among microorganisms is a relative phenomenon, meaning that high enough concentrations of an agent will generally kill even the so-called resistant strains. The simplest technique for isolating mutants resistant to antibiotics (or any other inhibiting drugs) is to plate large numbers of mutagenized cells on agar plates containing concentrations of the drug known to correspond to the first step resistance level. For example, sensitive strains of *E. coli* are known to experience mutations that afford them resistant to 10 μg/ml of ampicillin. By plating quantities of the wild strain onto agar containing 10 μg/ml, it is possible to isolate mutants that have acquired resistance to this drug.

Frequently the appropriate level of antibiotic to be used is not known, in which case the best attack is through the application of the gradient plate technique devised by Bryson and Szybalski. Petri plates are placed on a slanted surface so that when filled with a quantity of molten agar with antibiotic, a wedge-shaped layer is formed (Fig. 73). The concentration of antibiotic to use is somewhat arbitrary, being somewhere between twice and ten times the minimum inhibitory concentration for the wild strain. Several plates with different concentrations should be prepared. After the agar layer has solidified, the plates are placed in a horizontal position, and a second layer of agar growth medium without antibiotic is poured into them and allowed to solidify. The antibiotic in the lower layer, being a diffusible substance, will immediately begin to move into the upper layer. The result is the formation of a concentration gradient of the antibiotic from nearly zero at one end of a plate to approximately the original concentration of the lower layer at the other end.

A heavy suspension of the wild strain from which resistant mutants are sought is spread uniformly over the surface of the agar. Following twenty-four to forty-eight hours of incubation, a growth pattern similar to that shown in Figure 74 should result. If growth is confluent, the choice of antibiotic concentration was too low and should be increased by a factor of perhaps, ten. If no growth appears, the antibiotic concentration must be reduced by some factor, perhaps again ten. Obviously, some trial and error will be necessary, but if several concentrations are tried on the first preparation, success is usually possible.

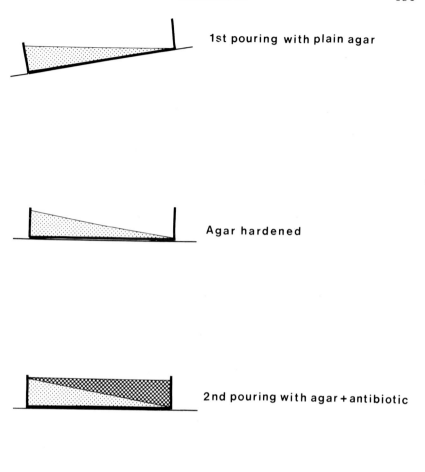

1st pouring with plain agar

Agar hardened

2nd pouring with agar + antibiotic

Gradient established

Figure 73. Preparation of a gradient plate. A Petri plate is first partially filled with molten nutrient agar while it is on an incline (top). After the agar has hardened, the plate is placed on a level surface and filled with molten agar that contains antibiotic or other substance. In a short time the antibiotic has diffused into the bottom layer and has established a concentration gradient. A heavy suspension of bacteria is spread over the surface of the agar and the plate is subjected to appropriate incubation. Typical results are seen in Figure 74.

Colonies that appear in areas of sparse growth are of possible
resistant mutants. Their identity must be confirmed to
eliminate the possibility of contaminants by staining and
biochemical tests. The tolerance level of the isolated mutants
can be determined, if required, by subjecting them to a serial
dilution assay in which the mutants are inoculated into a series
of antibiotic dilutions in broth that span the concentration
range used in the gradient plate for their isolation. One should
be aware that the position of a resistant colony on the gradient
plate has no reference to the mutant's maximum tolerance

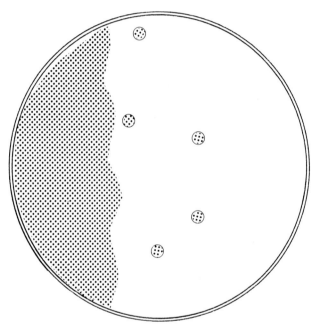

Figure 74. Results of the gradient plate technique. The
concentration gradient increases from left to right.
Confluent bacterial growth is observed in the area of low
antibiotic concentrations (left side) and ends where the
maximum tolerance for the wild strain is reached. Beyond
that point, in the area of higher concentrations, a few
colonies have formed, representing mutations that afforded their members tolerance toward higher concentrations.

level. It is sometimes possible to estimate a mutant's maximum tolerance by streaking the colony towards the high-concentration end of the plate and reincubating. Growth will develop along the streak up to the region corresponding to the cells' maximum tolerance. Figure 75 illustrates gradient plates that were subjected to this restreaking technique. Note that in plate B, the bulk of the growth from the streak stops abruptly, but a few colonies appear to have formed along the streak at

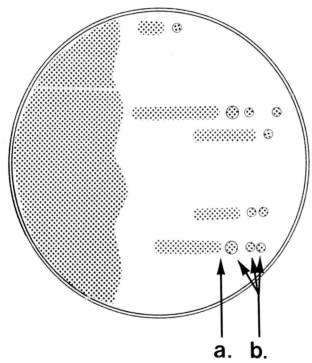

a. b.

Figure 75. Gradient plate following streaking of resistant colonies. The colonies that appeared in the area of high antibiotic concentrations (Fig. 74) are streaked toward yet higher antibiotic concentrations. Confluent growth ends abruptly where the maximum tolerance level of the first mutation is reached (a.). The few scattered colonies that appear beyond the confluent growth (b.) are made up of the progeny of cells that experienced a second mutation that afforded them still higher tolerance levels.

higher antibiotic concentrations. This pattern denotes a multistep process in the development of resistance towards the antibiotic used in plate B. The few colonies appearing in the high concentration region of the streaks represent subsequent mutations to higher tolerance levels. Resistance to the antibiotic used in plate A appears to exhibit the single-step pattern, however.

Radiation Mutants

We demonstrated earlier (Chapter Four) that exposure of a bacterial suspension to ultraviolet light of about 260 nm wavelength results in a logarithmic death curve (Fig. 49). It is possible to isolate from a culture of bacteria mutants that have acquired somewhat greater resistance towards the effects of UV radiation. In Figure 76 we have depicted UV survival curves for two strains of *E. coli,* one strain designated as resistant. Note that the resistant strain is still being killed by the UV light, but only after an initial delay, and at a slower rate. Strain A requires one minute of exposure to reduce its population by one log unit, while strain B needs 1.75 minutes to kill the same proportion of cells. Thus, as in the case of antibiotic resistance, mutationally acquired resistance to ultraviolet light also proves to be relative.

A second class of mutation observed in association with UV radiation is that which leads to greater sensitivity. That is, less radiation is required to kill the sensitive mutants compared with the wild strain. Most mutations that result in greater radiation sensitivity can be traced to one of many DNA repair mechanisms, some of which were discussed in Chapter Four. It stands to reason that repair mechanisms, presumably enzymatic systems, would be as susceptible to mutations as any other enzyme system of the cell. Nearly thirty sites have been identified on the *E. coli* chromosome, each of which, on experiencing a mutation, results in increased UV sensitivity. In many instances UV-sensitive mutants have lost the ability to support genetic recombination, inferring that at least some DNA repair may be closely related to or synonomous with the

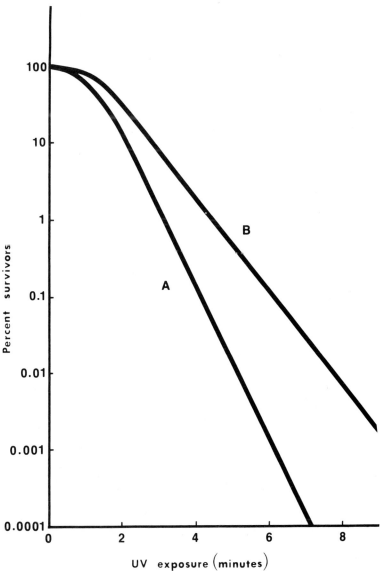

Figure 76. Sensitivity and resistance towards ultraviolet light. Two strains of *Escherichia coli* are tested for sensitivity towards UV. Curve A is the survival curve for the wild strain, whereas curve B depicts the survival of a UV-resistant mutant. The term resistance is therefore relative, in that the "resistant" strain is also killed by the radiation, but only at a slightly slower rate.

ability to support recombination. Other UV-sensitive mutants may not be able to effect *host cell reactivation,* the repair of UV-damaged phage DNA by the host cell. Still others appear to have lost the ability to support photoreactivation. Thus, the UV radiation sensitivity phenotype appears in reality to reflect the workings of a complex system of reactions involved with repair and replication of DNA.

Isolation

The parallel between antibiotic resistance and radiation resistance also applies to techniques for the isolation of the latter. A heavy dose of irradiation will kill most of the wild type cells in a culture, but in all probability will also kill most if not all of the few resistant mutants that may be in the culture as well. Evelyn Witkin devised a method whereby a small dose of UV radiation is directed at a bacterial culture, sufficient to kill 90 percent of the wild population. If one traces the effect of such a dose in Figure 76, it can be seen that the effect of such a dose on the resistant mutants is negligible. The culture is incubated for several hours, and then is exposed to a heavy dose of UV light, equal to fifteen times the earlier, low dose. This exposure kills all of the remaining wild cells, and many of the resistant cells, as well. The survivors, however, are all resistant mutants. The success of this technique is based partially on the behavior of the wild cells which, when irradiated, lose some ability to divide. As a result, their numbers remain low during the incubation period while the resistant mutants multiply normally. This method can be used to estimate the number of radiation resistant mutants in a bacterial population.

UV-sensitive Mutants

Use can be made of the characteristics associated with UV sensitivity, such as the loss of the abilities to support recombination, or to repair phage DNA, in the isolation of UV-sensitive mutants. Colonies of mutagenized bacteria are replica-plated onto agar plates that are heavily seeded with a second strain that is capable of undergoing recombination with the first (See Chapter Six for a discussion of bacterial recombination through conjugation). Those colonies that fail to demonstrate recombination probably belong to the recombinationless class of radiation sensitive mutants. The demonstration of recombination is easily brought about by using nutritional mutants in which case the replicated bacteria may be of the genotype $HfrA^-B^-C^+D^+$, and the recipient strain $A^+B^+C^-D^-$. The recipient bacteria are seeded on a minimal agar medium on which neither strain can grow, but recombiants of the genotype $A^+B^+C^+D^+$ can form visible colonies.

Colonies also may be replica-plated onto agar plates and irradiated with heavy doses of UV light. The colonies that fail to develop or show slight growth presumably were derived from sensitive bacterial mutants on the master plates that can be traced and picked.

A third technique involves infecting microcolonies of mutagenized bacteria with a low titer suspension of UV-irradiated bacteriophage. Those colonies that survive the infection have probably lost the ability to support host cell reactivation and are, therefore, possibly UV-sensitive mutants as well. As in all isolation procedures, the mutant genotype sought must be confirmed by appropriate means, which in the case of radiation mutants, involves the preparation of death curves as depicted in Figure 76.

Auxotrophs

Nearly half of the mutation sites mapped on the *E. coli* chromosome are associated with anabolic functions, such as the ability to synthesize an amino acid or vitamin. Mutations at these sites usually are manifested as a loss of the function, forcing the organism to rely on an exogenous source for the metabolite. An examination of the compositions of chemically defined media for the cultivation of various microorganisms reveals an immense range of nutritional abilities. The autotrophic bacteria require but a simple solution of a few inorganic salts, from which all of the organism's structural and energetic needs are derived (Table XIX).

TABLE XIX
MEDIUM FOR AUTOTROPHIC BACTERIA

K_2HPO_4	0.5 g
NH_4Cl	1.0 g
$CaSO_4$	1.0 g
$MgSO_4 \cdot 7H_2O$	2.0 g
Na Lactate (70%)	5.0 g
Salts solution*	50 ml (added after autoclaving)
Tap water	q. s. 1 L
*$FeSO_4 \cdot (NH_4)_2SO_4 \cdot 6H_2O$	1%

The familiar enteric bacterium *Escherichia coli* can grow on an inorganic salt medium with the addition of a single organic compound, glucose (Table XX). More fastidious organisms may require scores of preformed amino acids, vitamins, and other factors needed for growth (Table XXI). In those instances where an organism synthesizes one or more of its own metabolites, it can suffer a mutation at any time that blocks the

TABLE XX
MINIMAL MEDIUM FOR *ESCHERICHIA COLI*

NH_4Cl	1.0 g	Na_2HPO_4	6.0 g
$MgSO_4$	0.13 g	Glucose	4.0 g
KH_2PO_4	3.0 g	Distilled H_2O q. s. 1 L	

TABLE XXI
PARTIALLY DEFINED MEDIUM FOR
LACTOBACILLUS LEICHMANNII

Casein hydrolysate (vitamin free)	15.0 g	K_2HPO_4	1.0 g
Tomato juice	10.0 g	$MgSO_4$	0.02 g
Glucose	40.0 g	NaCl	0.02 g
Asparagine	0.2 g	$MnSO_4$	0.02 g
Na Acetate	20.0 g	Riboflavin	1.0 mg
Ascorbic acid	4.0 g	Thiamine	1.0 mg
Tween 80	2.0 g	Niacin	2.0 mg
L-cysteine	0.4 g	p-amino benzoic acid	2.0 mg
DL-tryptophan	0.02 g	Ca pantothenate	1.0 mg
Adenine sulfate	0.02 g	Pyridoxine	4.0 mg
Guanine hydrochloride	0.02 g	Folic acid	0.2 mg
Xanthine	0.02 g	Biotin	0.008 mg
Uracil	0.02	Vitamin B12	0.00003 mg
KH_2PO_4	1.0 g	Distilled water q. s. 1 L	

formation of one of the metabolites. If the particular metabolite is then not available to the organism in the medium, the mutant cells are unable to continue growth. Such mutants that require one or more growth factors in the medium not required by the parent cells are called *auxotrophs;* the parent cells are then referred to as *prototrophs.* The auxotrophs have proven to be an immensely valuable class of mutants. Auxotrophs of *Neurospora* played a major role in the establishment of the One gene-One enzyme hypothesis of Beadle and Tatum.

In recent times, auxotrophs have been responsible for the elucidation of dozens of metabolic pathways. Many mutations that lead to auxotrophy prove to have resulted in the loss of the formation of an enzyme involved in some step leading to the required metabolite. A fortuitous characteristic of many auxotrophic mutants is the ability to satisfy a specific metabolite loss by utilizing certain sequent metabolites. Thus, in the formation of biotin in *E. coli* (Fig. 77), while the organism is prevented from forming the vitamin due to a mutation that blocks the step, for example from pimelic acid to 7-keto-8-aminopelargonic acid, growth can be supported if any of the intermediates "downstream" of the blocked step are added to the medium, such as 7-keto-8-aminopelargonic acid, 7, 8-diaminopelargonic acid, or desthiobiotin, as well as biotin itself. This ability is the key to using auxotrophs in the determinations of metabolic pathways.

Suppose in our example of biotin synthesis we did not know the identity of the precursors involved. In the laboratory we would isolate a number of biotin auxotrophs, and then supply them with several compounds that *could possibly* act as precursors to biotin. The choice of compounds is based on knowledge of biochemistry, and a certain amount of intuition. Eventually a growth-support pattern similar to that depicted in Table XXII develops. If the collection of mutants isolated fairly well represents all of the immediate steps involved in the formation of biotin, then one can begin to identify the intermediates involved and arrange them in the order in which they participate in the pathway. We see that the growth of all mutant strains is supported by biotin itself, whereas none grow

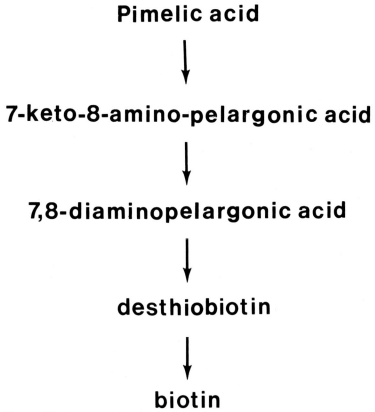

Pimelic acid

↓

7-keto-8-amino-pelargonic acid

↓

7,8-diaminopelargonic acid

↓

desthiobiotin

↓

biotin

Figure 77. Proposed biosynthetic pathway to biotin, based on data derived from studies of biotin auxotrophs (see Table XXII). (From Pai, C. H.: *Canad J Microbiol, 15:* 21, 1969).

in the presence of pimelic acid, diaminopimelic acid, or lysine. The compounds can be arranged according to the number of mutants showing growth in their presence. The compounds that support the greatest number of mutants are probably nearest to biotin in the pathway; those that support the fewest number are farthest from biotin. This is because in a random collection of auxotrophs, there is a greater probability that a given metabolite will support the growth of an auxotroph the nearer its blocked step is found to the end of the pathway. 7-keto-8-aminopelargonic acid supports the growth of three of the mutants, 7, 8-diaminopelargonic acid supports seven, and desthiobiotin supports nine of the twelve mutants. This is, therefore, the order in which these metabolites participate in the formation of biotin.

TABLE XXII

GROWTH OF BIOTIN AUXOTROPHS IN PRESENCE OF VARIOUS POSSIBLE PRECURSORS OF BIOTIN*

Mutants			*Compounds Tested*			
	d-biotin	*PA*	*7,8DAPA*	*7A8KPA*	*dl-desthio-biotin*	*7K8APA*
I	+	–	–	–	–	–
II	+	–	–	–	+	–
III	+	–	+	–	+	–
IV	+	–	+	–	+	+

⁺ = growth in minimal medium in presence of added compound
⁻ = no growth in minimal medium in presence of added compound
PA = pimelic acid
7,8DAPA = 7,8-diaminopelargonic acid
7A8KPA = 7-amino-8-ketopelargonic acid
7K8APA = 7-keto-8-aminopelargonic acid

*From Pai, C. H.: *Canad J Microbiol, 15:*21, 1969. Reproduced by permission of the National Research Council of Canada.

There is no evidence from this experiment that pimelic acid participates in the formation of biotin, but from other independent studies it does appear to act as a precursor for 7-keto-8-aminopelargonic acid. It should be emphasized that such a scheme as is shown in Figure 77 that is derived from the study of auxotrophs is only tentative. There may be further intermediate steps that are undetected. The putative metabolites tested may not necessarily participate directly in the pathway, but may be altered beforehand. Confirmatory enzymological and genetic experiments must be carried out. One such confirmatory test involves a phenomenon known as *syntrophism,* or *cross-feeding.*

Syntrophism

Organisms that have suffered a block in a biosynthetic pathway as a result of a mutation very frequently accumulate and excrete the metabolite immediately before the blocked step, provided no control mechanism is triggered to stop its formation. In a generalized example (Fig. 78), mutants I, II, and III are unable to grow alone in minimal media lacking compound A. They will grow in minimal media if in close proximity, however. Mutant I cannot convert compound B to A, the blocked step resulting in the accumulation of B in the medium. Mutant III cannot convert C to B, but it can take up the B excreted by mutant I and convert it to A. A small quantity of A is then released into the medium through normal secretion, leakage, and the occasional death of mutants II and III to support moderate growth of mutant I.

Syntrophism cannot be demonstrated in every case of auxotrophy, mainly due to control mechanisms or permeability conditions that prevent the release of intermediates, but in those instances where it is possible, one can deduce the order in which a collection of auxotrophs have suffered mutations in the pathway even before the intermediate compounds are tested.

Isolation

The isolation of auxotrophic mutants presents difficulties not present in the straight-forward isolation of antibiotic resistant mutants. In the latter instance, the antibiotic is the selective agent, the presence of which selects *for* the mutants and *against* the wild type. Not so in the isolation of auxotrophs, for auxotrophy is at best genetically neutral for the mutants in a rich medium, and definitely negative in a medium lacking the required nutrient. In order to isolate auxotrophs efficiently the situation must be reversed so that the mutants have the advantage over the wild type. This is actually the case in the penicillin technique devised independently by Davis and by Lederberg for the isolation of auxotrophs of *E. coli*.

The technique is based on the phenomenon of penicillin acting only on actively growing cells. Thus, a suitably mutagenized suspension of *E. coli* in a minimal medium (Table XX) but lacking NH_4^+ is subjected to a brief starvation period to rid the cells of stored nutrilites, then penicillin and

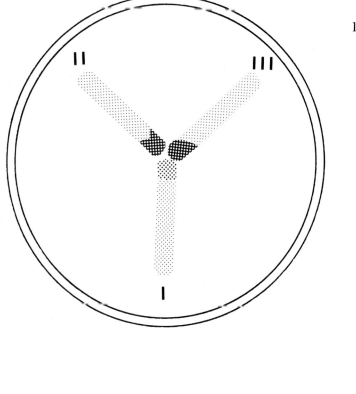

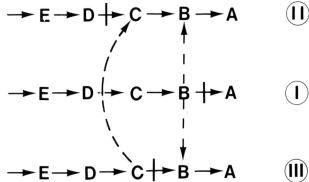

Figure 78. Example of syntrophism. Three auxotrophic mutants lacking compound A are streaked on a minimal medium lightly supplemented with compound A. Sparse growth is observed except where the strains are in close proximity. Strains II and III show heavy growth where strain I has released excess amounts of compound B into the medium as a result of the block in the conversion of B to A. At the same time, strain III releases some compound C into the medium as a result of the block it suffered in the step from C to B, stimulating the growth of II as well. A small amount of compound A released by strains II and III supports moderate growth of strain I.

(NH₄)₂SO₄ are added and incubation is carried out for three hours (Fig. 79). The prototrophic parental cells are capable of forming all of their nutritional needs from the constituents of the medium and thus begin to grow. Any auxotrophs in the suspension will lack one or more metabolites and will not grow. The penicillin will only affect the growing parental cells, killing them and sparing the auxotrophs. Following the penicillin treatment, penicillinase is added to destroy the antibiotic, and the suspension is plated onto a rich medium. Most of the surviving cells should be auxotrophs. Auxotrophy is confirmed by replicating the colonies that develop onto minimal agar medium. Those that grow on rich medium but not minimal are considered possible auxotrophic mutants.

The determination of what specific metabolite the mutants lack can be carried out by plating the mutants onto minimal media that contain pooled mixtures of selected metabolites (See Clowes and Hayes for details). If one is interested in isolating mutants auxotrophic for specific metabolites, such as tryptophan or paraaminobenzoic acid, the growth factor may be added to the minimal medium following the penicillinase step, and the mixture incubated. Those mutants that can use the growth factor will increase in number and their isolation becomes considerably more probable in later steps. This technique may have pitfalls in that many bacteria are capable of various metabolite interconversions. An organism may have lost its ability to synthesize compound A from B, but given C, it may be able to convert C directly to A, appearing as an auxotroph lacking C. It is thus imperative that all auxotrophs isolated are carefully tested to confirm their mutant phenotype.

In instances where a given bacterium is not susceptible to the penicillin techniques, one can go directly to replica plating. Cells are treated with mutagen and then plated on complete media at dilutions to insure isolated colonies. Following incubation, the developed colonies are replica plated onto minimal agar plates and these are incubated. The minimal agar plates are then examined side-by-side with each of their matching master plates, and those colonies that appear on the complete medium but not on the minimal agar probably

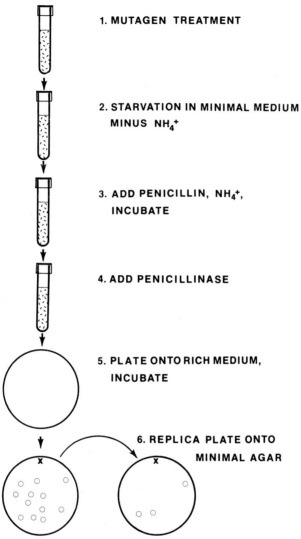

1. MUTAGEN TREATMENT

2. STARVATION IN MINIMAL MEDIUM
   MINUS NH₄⁺

3. ADD PENICILLIN, NH₄⁺,
   INCUBATE

4. ADD PENICILLINASE

5. PLATE ONTO RICH MEDIUM,
   INCUBATE

6. REPLICA PLATE ONTO
   MINIMAL AGAR

Figure 79. Penicillin method for the isolation of auxotrophic mutants of *E. coli*. A suspension of bacteria is treated with an appropriate mutagen (1) after which the cells are suspended in a minimal medium less a nitrogen source for a brief period (1 to 3 hours) to rid them of stored nutrilites (2). Penicillin and an inorganic nitrogen source are added and incubation is continued for an additional 3 hours. Penicillinase is added to destroy residual antibiotic, followed by the spreading of undiluted aliquots onto a rich medium (5). The few colonies that develop on the agar plates following incubation are presumed to be auxotrophs, and as confirmation, are replica plated onto minimal agar. Clones that grow on the minimal agar contain wild type cells that escaped the effects of the penicillin and are eliminated from consideration; all others are carried through further tests to determine the nature of the auxotrophy.

represent auxotrophic mutants. Auxotrophy is then confirmed by transferring suspected colonies onto minimal agar slants, and specific lost metabolites are identified as described above.

Fermentation

The capacity of many bacteria to utilize various carbohydrates as sources of carbon and energy can be lost through a single mutation. Thus, while fermentation reactions are of enormous value in the identification and classification of bacteria, they are subject to the same genetic influences as other functions of the cell. *E. coli,* noted for its typical acid reaction in lactose-containing differential media, experiences spontaneous mutations that bring about the loss of this characteristic.

Isolation

If one plates several thousand isolated wild *E. coli* cells onto lactose eosin methylene blue (EMB) agar plates, a few developing colonies will show the typical lac⁻ pink morphology (Fig. 80). A more efficient method for isolating lac⁻ mutants is to incubate a culture of the wild type with a compound that is converted to a toxic derivative by lac⁺ cells. In the isolation of lactose-permease minus mutants, o-nitrophenyl-β-D-thiogalactoside is included in the medium. Wild cells take up this compound and are severely inhibited, while lac⁻ mutants lacking the permease cannot take it up and therefore grow normally.

The isolation of fermentation back-mutants (lac⁻——→lac⁺) is considerably more efficient in that one merely has to plate a heavy suspension of lac⁻ cells onto lactose EMB agar. Confluent pink lac⁻ growth will result, but occasional lac⁺ clones will show up as dark red papillae against the light background. The papillae are picked and restreaked for pure culture isolation. Techniques with EMB agar are not confined to using lactose. Any carbohydrate that results in the formation of sufficient

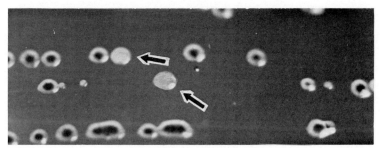

Figure 80. Lac⁻ colonies on EMB agar.

acid on EMB agar can be used. Other differential fermentation media such as tetrazolium can also be applied to the isolation of fermentation mutants.

Bacteriophage Resistance

Bacterial viruses begin an infectious cycle by attaching themselves to specific adsorption sites on the bacterial host cell surface. A susceptible bacterium can experience a mutation that alters the structure of the adsorption sites such that the particular phage can no longer attach to the cell. The bacterium thus exhibits resistance towards that bacteriophage. We emphasize the point that bacterial mutations that lead to phage resistance are always directed towards a *specific* phage or class of phages. Mutations that afford *E. coli* resistance against infection by T₄ phage do not protect this bacterium against T₁, lambda, or φX174, for example.

Isolation

Bacteriophage-resistance mutations are highly selective, making isolation of corresponding mutant cells relatively easy. Heavy suspensions of susceptible bacteria and appropriate phage are mixed and plated on "soft agar" plates (see Chapter Seven). Most of the bacteria obviously will be lysed, but a few isolated colonies will appear following suitable incubation. The colonies are picked, purified, and confirmed as being phage resistant. The customary designation of phage resistant *E. coli* mutants has been to indicate the phage type towards which the bacterial mutant is resistant following a slash sign. Thus, *E. coli* B/2 denotes strain B resistant to T₂ phage.

Motility and Flagella

Certain bacteria are capable of vigorous movement in liquid or semisolid media through the action of long, wavy, filamentous appendages known as flagella (Fig. 81a). Flagella are constructed of repeated homogenous subunits of a protein called *flagellin* that has a molecular weight of about 2 to 4 x 10⁴. The point of attachment of the flagella to the cell membrane is a complex apparatus consisting of a hook, discs, a basal body, and other parts, each of which is chemically distinct from one another and from *flagellin*. This apparatus is probably responsible for movement of the flagella as well. Thus, a number of genes presumably are involved with flagellar structure and function, and in fact at least fifteen genes associated with motility have been identified in *Salmonella,* and nearly a like number in *E. coli.*

Mutations in flagellar structural genes run the familiar gamut of expressions, from slight modifications of filament structure, with or without effect on normal motility, to complete loss of the filament structure or one of the components of the attachment apparatus.

One of the most striking consequences of mutations in flagellar structural genes results in the so-called *curly* phenotype (Fig. 81b). A single amino acid substitution in the flagellin molecule causes a halving of the flagellum's wavelength. Curly mutants show spinning motility in liquid media but fail to swarm on semisolid motility agar. Other curly mutants possess flagella with a greater amplitude as well as a smaller wavelength, giving them a hooked appearance. Some curly mutants form flagella that are part curly and part normal.

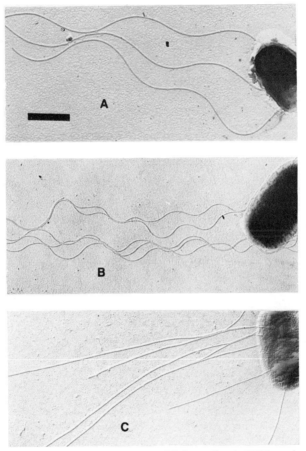

Figure 81. Flagellar mutants of *Salmonella*. A. Wild strain with normal flagella. B. Mutant strain showing curly flagella. C. Mutant strain showing straight flagella. Cells were fixed with osmic acid and shadowed with chromium. Bar indicates 1 micrometer. (Electron micrographs kindly supplied by T. Iino.).

One structural gene mutation leads to the flagellar phenotype known as *straight,* which obviously denotes a loss of the wavy feature of flagella. Straight mutants are nonmotile (Fig. 81c).

Mutations in the *mot* genetic region create a condition of paralysis, in which the cells may form normally appearing flagella but are incapable of movement. The basis for the *mot* mutant phenotype is not known, but it may involve the motility apparatus itself or its energy source.

Some years ago (1922) it was discovered that *Salmonella* exhibited a peculiar phenomenon called *phase variation*. A pure strain of *Salmonella typhimurium* would dissociate into two populations possessing antigenically distinct flagella. The bacteria appeared to alternate between the two antigenic phases in a periodic manner suggestive of a series of forward and back mutations, although with a rapidity somewhat greater than other mutation rates were known to express. Detailed analysis of the genetic regions responsible for motility in the *Salmonella* have now revealed a remarkable, coordinated system of regulatory and structural genes. It is the interaction of the components of this system that is responsible for phase variation; mutations principally are not involved.

Two structural genes in *Salmonella* code for the formation of the two antigenically distinct flagellar proteins. They are designated H1 and H2 and are controlled by two regulatory genes, ah1, and ah2, respectively. The latter genes may be analogous to operator regions discussed in Chapter Three. A third class of gene involved in this system is designated vh2. It has been suggested that the vh2 gene acts as a "clock" in bringing about the periodic repression and derepression of the ah2 gene. In the repressed state, the ah2 region switches off the H2 structural gene, freeing the H1 gene to make phase 1 flagella. Gene ah2 in the derepressed state inactivates the ah1 gene, activates the H2 gene and phase 2 flagella are then made. The nature of the vh2 clock mechanism has been thought of as merely one of instability due to some chromosomal structural anomaly, the state of instability being reflected in the observed alternating phase variation.

By recording the phenotypes that mutations in these genes exhibit, it has been possible to formulate a picture of how these genes operate. Mutations in either the H1 or H2 genes lead to structural alterations, such as the curly phenotype. Mutations in the ah1 or ah2 genes frequently result in the mutant becoming monophasic for the flagellar type not affected by the mutation.

That is, should an organism suffer a mutation in the ah2 gene, it becomes monophasic type 1. If the vh2 region experiences a mutation, the organism becomes monophasic for the type it happened to be expressing at the time of the mutation.

Another genetic site involved in motility is the *fla* region, which probably consists of several distinct genes. These appear to be regulatory genes having control over various aspects of flagellar synthesis, perhaps the flagella-forming machinery itself. Mutations in the *fla* region may result in either no flagellin being formed, or measurable quantities being formed but with no assembly of the filament structure.

Isolation

Since most flagellar mutants exhibit reduced motility, their isolation is normally carried out on motility agar plates on which mutagenized cells are distributed for isolated colonies. In one technique, cells are spread over a 1.5% agar surface and then overlayed with 0.4% agar. Colonies that fail to show typical flaring, or show a notable reduction in colony size, are suspected motility mutants and can be picked and tested further. Another technique takes advantage of a class of *Salmonella* bacteriophage that are *flagellotropic;* that is, they initiate infection by attachment to flagella. Mutants lacking flagella are therefore resistant and can be easily isolated.

Minicells

Recently a mutant strain of *E. coli* was discovered that exhibited an aberrent, asymmetric mode of cell division. On occasions cell division would occur very near one end of the cell, resulting in one daughter cell being small, spherical, and anucleate (Fig. 82). In addition to lacking chromosomal DNA, the *minicells,* as they are called, also are without certain enzymes associated with DNA, such as DNA-dependent RNA polymerase, DNA methylase, and the photoreactivating enzyme. Some DNA polymerase activity is still detectable in the minicells, and levels of RNA and protein are near those in normal cells. DNA-ligase activity is also at a normal level. While minicells do not contain parent chromosomal DNA, they may contain the DNA of any plasmids carried by the parent cells. Plasmids, which are self-replicating extrachromosomal genetic elements (Chapter Seven) not only are capable of replication in minicells, but also can direct the synthesis of RNA and protein when supplied with precursor molecules. Another observation made of minicells is that these bodies are susceptible to coliphage infection, in which small numbers of mature, infectious phage particles are produced. Since minicells appear to support DNA, RNA, and protein synthesis in the absence of the influence of parent bacterial DNA, these mutants offer a remarkable system with which to study these functions.

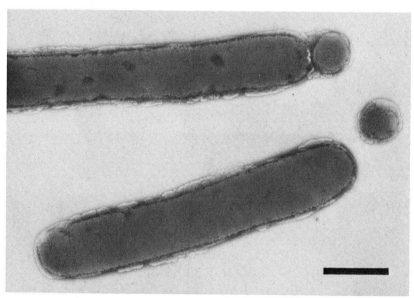

Figure 82. Minicells of *Escherichia coli.* Shown here are a normal cell (a.), a minicell (b.), and a cell undergoing polar division which will result in the production of an anucleate minicell (c.). Preparation negatively stained with phosphotungstic acid. Bar represents one micrometer. (Micrograph courtesy of H. I. Adler and D. P. Allison, Biology Division, Oak Ridge National Laboratory)

Isolation

Minicells were fortuitously discovered in a culture of irradiated *E. coli* strain K-12 cells. Minicells occur in mature cultures of the mutant strain in numbers roughly equal to those of the parents. Their separation from the parent cells is accomplished by sucrose density gradient centrifugation, minicells being somewhat less dense. An initial screening step with penicillin has been shown to increase minicell yield.

Sporulation

The formation of endospores is found essentially among members of the genera *Bacillus* and *Clostridium*. In *Bacillus subtilis* at least twenty genes, randomly scattered throughout the chromosome, have been identified with sporulation, and another thirty are thought to be involved. Mutations in any one such gene can prevent the formation of mature spores. Microscopic examination of various asporogenous mutants reveals that spore formation can be blocked at different stages of development. Some mutants show no signs of any development, while others appear to have reached a relatively advanced stage of prespore formation.

Isolation of asporogenous mutants of *B. subtilis* is based on the fact that the mutants frequently exhibit colonial morphologies different from the wild, spore-forming type. The colonies of asporogenous mutants are usually larger and more translucent than those of the parent strain, or frequently show less pigmentation after prolonged incubation.

INDUSTRIAL CONSIDERATIONS

In Chapter Three we discussed the practice of intervening in microbial biosynthetic control systems for the purpose of increasing industrial fermentation yields. Subjecting microorganisms to mutagenesis and selection also has been an approach to that end. In the production of antibiotics such as penicillin, tetracycline, and streptomycin, repeated physical and chemical mutagenic treatment and selection of specific strains have been responsible for ultimately increasing yields by factors of 1000 percent or more. Improvements in product yield of 20 to 30 percent are more commonly encountered, but from an economic standpoint are still considered highly desirable. On a practical basis, increases of less than about 10 percent are difficult to detect by routine screening procedures.

Mutagenesis usually is carried out on spore suspensions of promising fungal, actinomycete, or bacterial strains, and screening is done by spotting the mutagenized spores onto pour plates containing organisms susceptible to the antibiotic under consideration. The selection of the optimal dose of mutagen appears critical in the induction of improved industrial stains. It has been shown (Fig. 83) that as one increases

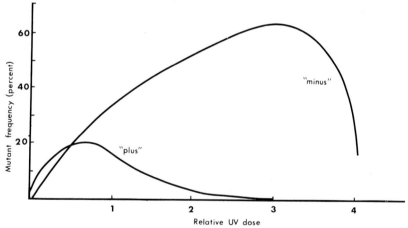

Figure 83. Effect of varying doses of ultraviolet light on the frequencies of "plus" and "minus" mutants. An actinomycete is irradiated with UV to produce mutants with improved antibiotic production ("plus" mutants). At the same time, "minus" mutants appear, exhibiting lower yields than the wild type. As dose is increased, the proportion of "plus" to "minus" mutants remains close to unity, until a certain dose level is reached. At this point, "minus" mutants continue to accumulate, whereas the frequency of the more desirable "plus" mutants declines to near zero. (From Alikhanian, S. I.: *Adv Appl Microbiol, 4*:1, 1962)

mutagen dose, the proportion of desirable ("plus") mutants rises to a peak and then drops. A further increase in dose results in an increase in the number of undesirable ("minus") mutants. Plus and minus refers to the *specific yield* of the desired product as being greater than or less than that of the wild type, respectively. Obviously, the dose that corresponds to the peak of the plus strain must be determined, or at least its existence must be recognized and several dose levels (and several mutagens, as well) should be tried.

Initial screening plates are appropriately incubated and zones of inhibition around isolated colonies are measured and compared with those of the wild strains (Fig. 84). If a growth factor such as a vitamin is sought rather than an antibiotic, then the test organism suspended in the pour plates is one that requires the growth factor and zones of stimulation are measured. A screening procedure such as this is costly with respect to manpower and materials, but since the production of large quantities of antibiotics and other commercially desirable compounds by microorganisms usually offers them little selective advantage in pure cultures, the cost has to be paid.

A further lesson the industrial microbiologist must learn is that results obtained from laboratory cultures do not necessarily reflect the behavior of a given microbial strain growing in a 100,000 liter fermentor. Once an isolated strain shows commercial promise, its producing characteristics are carefully followed in gradual stages from laboratory scale (a few hundred milliliters to a few liters) to pilot plant scale (ten to a hundred liters), and finally to plant scale (up to 100,000 liters or more). Every step is monitored as to yield, pH, oxygen tension, temperature, agitation rate, and other factors.

An actual increase in the production of a specific compound is not necessarily the only trait considered desirable in improved strains of industrial microorganisms. An organism may produce, in addition to an antibiotic, other compounds that interfere with the economic isolation and purification of the drug. Strains are sought that produce lesser amounts of the interfering compounds, and in fact it has been the case that an "improved" strain may produce slightly less quantities of the

Figure 84. Antibiotic screening plate. Several strains of *Penicillium* are screened for antibiotic production simultaneously. Once fungal growth is established (24 to 48 hours), the plate is flooded with a suspension of a suitable indicator organism, in this case *Staphylococcus aureus,* and reincubated. Zones of inhibition around fungal colonies indicate relative antibiotic production for each of the strains.

antibiotic, but the absence of the interfering material results in a greater overall yield or a purer product.

It may be required to find strains of antibiotic-producing microorganisms that produce compounds of broader antimicrobial spectra than those of the wild strains. Frequently broader antibiotic activity can be traced to a mutation-induced alteration in the molecular structure of an antibiotic.

Patentability of Industrial Microbial Strains

It has generally been the custom of the patent offices of the United States and most major foreign countries (Japan, England, Germany) to issue patents on specific strains of microorganisms that have some commercial value. In order to patent a microbial strain, certain criteria must be met. First, the strain must be truly new and novel. A description of the strain cannot have been published in the open literature prior to application for the patent. The publication requirement is strictly adhered to in most foreign countries, but the United States Patent Office allows up to one year from the time a description is published until patent application is made. A subculture of the strain must be deposited in a recognized culture collection such as those listed in this chapter prior to the filing of the patent application. Enough of a description of the strain must appear in the patent application to distinguish it from other strains of the same species. At least one nonfunctional characteristic must be included. A nonfunctional characteristic is one that is unrelated to the function for which the organism is primarily being registered. Figure 85 is a copy of a typical United States patent on a *Streptomyces* strain that produces streptomycin.

Patented Oct. 16, 1951   **2,571,693**

UNITED STATES PATENT OFFICE

2,571,693

PROCESS FOR PRODUCTION OF STREPTOMYCIN

Eugene L. Dulaney, Rahway, N. J., assignor to Merck & Co., Inc., Rahway, N. J., a corporation of New Jersey

No Drawing. Application July 25, 1950, Serial No. 175,891

1 Claim. (Cl. 195—80)

This invention relates to the production of streptomycin, and more particularly, to procedures for obtaining streptomycin in enhanced yield by the utilization of the new and distinct culture of *Streptomyces griseus* hereinafter described and characterized.

In my pending application, Serial No. 8,308, now U. S. Patent 2,545,572, filed February 14, 1948, I have disclosed procedures for producing streptomycin in yields of the order of 800 to 1100 mcg./ml. by means of the new and distinct culture *S. griseus* Dulaney L-118. This culture was obtained by subjecting spores of a strain of *S. griseus* characterized as resistant to an initial streptomycin concentration of at least 500 mcg./ml. to the action of ultra-violet light, allowing the spores surviving such treatment to grow and selecting from the resulting colonies of *S. griseus* those exhibiting an increased ability to produce streptomycin, again subjecting spores from the selected colonies to the action of ultra-violet light and repeating the colony development, selection, and spore irradiation until a mutated strain of *S. griseus* was obtained which possessed the capacity of consistently producing streptomycin in yields of at least 800 mcg./ml.

While the streptomycin yields of 800 to 1100 mcg./ml. produced by the culture *S. griseus* Dulaney L-118 are markedly superior to yields obtainable with strains of *S. griseus* previously available, I have now discovered that yields of streptomycin can be further increased to about 1800 to 2200 mcg./ml. by utilizing the new and distinct culture hereinafter designated as *S. griseus,* albus mutant (Dulaney Z-38). This new culture of mutant was obtained by subjecting spores of *S. griseus* Dulaney L-118 to the mutating action of ultra-violet light and soft X-rays. The experimental procedure leading to the isolation of the new mutant was as follows:

Spores of *S. griseus* Dulaney L-118 were washed from a surface growth with sterile distilled water, and the resulting spore suspension was filtered through several layers of sterile absorbent cotton. The filtered spore suspension was then exposed to ultra-violet light having a wave length of 2,537 Å for sufficient time, about 2 minutes, to kill approximately 99% of the spores. The treated cell suspension was then diluted with distilled water, and a quantity of the diluted suspension was spread over Petri dish plates of nutrient agar, i.e., an aqueous medium containing 2.5% glucose, 4.0% soybean meal, 0.25% sodium chloride, 0.5% distillers solubles and solidified with agar. After incubation until good growth and sporulation occurred, the colonies were transferred to agar slants of the same medium, and, after incubation until good growth and sporulation occurred, sterile water was added to each of the slants, and separate spore suspensions were prepared from each slant.

Quantities of each spore suspension were transferred to flasks containing an inoculum medium composed of 1% glucose, 1% enzymatic digest of casein, 1% sodium chloride, 0.6% meat extract, and distilled water to volume. The inoculated flasks were then incubated with constant agitation for 48 hours, and the vegetative growth which developed was used to inoculate flasks of aqueous fermentation medium having the composition: 2.5% glucose, 4.0% soybean meal, 0.25% sodium chloride and 0.5% distillers solubles. These inoculated flasks were then incubated at 28° C. with constant agitation for 4 to 5 days to permit maximum streptomycin production.

The foregoing mutation and selection procedures were repeated with additional spore suspensions of *S. griseus* Dulaney L-118 until a superior streptomycin producing mutant was obtained as evidenced by an enhanced yield of streptomycin in the fermentation flask.

A spore suspension in distilled water of this superior mutant which was designated O-3 was exposed to soft X-rays until approximately 99% of the spores were killed. This treated suspension was diluted with distilled water, plated on nutrient agar for spore development, and the resulting spores were used to inoculate flasks of inoculum medium and flasks of fermentation medium in accordance with the procedure above-described.

A new and higher yielding mutant thus obtained from O-3 was designated R-315. Spores of the mutant R-315 were similarly treated with X-rays, and from the cultures thus obtained mutant T-535 was selected as a parent for further mutation work. Spores of mutant T-535 were treated with ultra-violet light and mutant V-148 was selected from the resulting cultures. One of the natural isolates of mutant V-148, i. e., strain X-69, was found to be a superior streptomycin producer and was, in turn, treated with ultra-violet light yielding the superior mutant Z-38, more fully hereinafter referred to as *S. griseus,* albus mutant (Dulaney Z-38). The following diagram illustrates the genealogy of strain Z-38.

```
              ultra-violet     X-rays         X-rays
L-118      ─────────▶   O-3 ─────────▶  R-315 ─────────▶ T-535
              light                 ultra-violet              │
                                                              │
         ┌──────────────────────── ◀────────── ◀─────────────┘
         │                         light
         ▼                     ultra-violet
V-148      ─────────▶   X-69 ─────────▶  Z-38
              natural               light
              isolate
```

Comparative morphological properties and biochemical reactions as given in Bergey's Manual of Determinative Bacteriology and as observed for *S. griseus* Dulaney L-118 and *S. griseus,* albus mutant (Dulaney Z-38) are tabulated below:

CULTURAL CHARACTERISTICS OF *S. GRISEUS*

	Bergey's Manual of Determinative Bacteriology	*S. griseus* L-118	*S. griseus* Z-38
Filaments	Branching, a few spirals	Straight, branching, no spirals	Straight, branching, occasional spirals.
Conidia	Rod-shaped to short cylindrical 0.8 x 0.8 to 1.7 microns.	Agrees	Agrees.
Gelation stab	Greenish-yellow or cream-colored surface growth, brownish tinge, rapid liquefaction.	16 days—50% of medium liquefied (color not recorded).	16 days—25% of medium liquefied (color not recorded).
Synthetic Agar	Thin, colorless, spreading; olive buff-aerial mycelium thick, powder, water-green	Colorless growth no aerial mycelium.	Colorless growth no aerial mycelium.
Starch Agar	Thin spreading, Transparent	Thin, spreading, not transparent, cream growth, starch hydrolyzed.	Thin, spreading, not transparent, cream growth, starch hydrolyzed.
Dextrose Agar	Elevated in center, radiate cream-colored to orange, erose margin. Abundant, cream-colored, almost transparent.	Faint growth-spreading, colorless	Faint growth-spreading, colorless.
Plain Agar	Abundant, yellowish pellicle with greenish tinge, much folded.	Colorless, transparent growth	Colorless, transparent growth.
Dextrose broth	Cream colored ring, coagulated with rapid peptonization, becoming alkaline.	Pellicle (Color and type not observed).	Pellicle (Color and type not observed).
Litmus milk	Yellowish wrinkled	Peptonized (pH of substrate, and color and type of surface growth not observed).	Peptonized (pH of substrate, and color and type of surface growth not observed).
Potato		Heavy growth, grey; rugose; potato darkening.	Heavy growth, tan; rugose, potato darkening.
Reduction	Nitrites produced from nitrates	Nitrites produced	Nitrites produced.
Pigment	Not soluble	Not soluble	Not soluble.
Oxygen tension	Aerobic	Aerobic	Aerobic.

Additional typing tests not listed by Bergey gave the following results:

Test	S. griseus L-118	S. griseus Z-38
Cellulose decomposition	No decomposition	No decomposition.
Ca malate agar	Faint growth	Faint growth.
Tyrosine agar	No apparent growth	No apparent growth.
Phosphate agar	Spore-bearing hyphae in clusters.	Vegetative mycelium.
Soybean agar (Composition, see page 2, lines 22-24).	Abundant sporulation, grey to grey-green spores.	Abundant sporulation, spores white.
1% Yeast extract, 0.5% glucose agar.	Abundant sporulation.	No sporulation.
Spore inoculum[1] on Difco® Yeast Beef Agar (No. B244).[2]	Light sporulation.[4]	Practically no sporulation.[4]
Vegetative inoculum[3] on Difco Yeast Beef Agar (No. B244).	About 80% of colonies show sporulation.[3]	Less than 1% of colonies show sporulation.[3]

[1] Spores from agar Blake bottle suspended in distilled water as previously described.
[2] The composition of Difco Yeast Beef Agar (No. B244) is as follows:

Ingredient	Grams per 100 cc. of aqueous medium
Bacto® Beef Extract	0.15
Bacto Yeast Extract	0.3
Bacto Peptone	0.6
Bacto Dextrose	0.1
Bacto Agar	1.5

NOTE: To this medium was added 1 gram per 100 cc. of Difco dehydrated agar to yield a firmer solidified medium.

[3] Vegetative inoculum was prepared by inoculating 40 cc. of a sterile liquid medium described below, in a cotton plugged 250 cc. Erlenmeyer flask and incubating at 27° C. for 20 hours on a rotary shaker.

The medium used for vegetative inoculum development had the following composition:

Ingredient	Grams per 100 cc. of aquerous medium
N-Z-Amine (enzyme hydrolized casein)	1.0
Dextrose	1.0
Beef Extract	0.3

[4] In inoculating with spores, spores from a Blake bottle culture were suspended in sterile water at a sufficient dilution so that when loop streaked on the test medium, and incubated at 27° C. for about 6 days, separate colonies developed and were examined individually for sporulation.

[5] A loop of the vegetative inoculum, prepared as described in note (3) above, was streaked on the test medium. Dilution was such that, after incubation of the test plate for about 6 days at 27° C., separate colonies developed and were examined individually for sporulation.

The following examples are presented to illustrate procedures for the production of streptomycin using the new culture *S. griseus,* albus mutant (Dulaney Z-38).

EXAMPLE 1

Spores from nutrient agar slants of the *S. griseus* strains L-118, X-69 and Z-38 previously referred to were used to inoculate separate flasks containing the following nutrient medium: 1% glucose, 1% enzymatic digest of casein, 1% sodium chloride, 0.6% meat extract and distilled water to volume. The flasks containing the inoculated medium were incubated on a rotary shaker at 28° C. until good growth occurred. After 48 hours incubation, this vegetative growth was used to inoculate 250 ml. Erlenmeyer flasks containing 40 ml. of the following medium:

Glucose . . . . .per cent   2.5
Soybean meal . . . . .do   4.0
Sodium chloride . . . . .do   0.25
Distillers dried solubles . . . . .do   0.5
Distilled water . . . . .to volume
pH before sterilization 7.38

The flasks containing the inoculated medium were incubated on a rotary shaker, 220 R. P. M., at 28° C. until maximum streptomycin production, as measured by the *Bacillus subtilus* cup assay, occurred. The results of the comparative experiment are given in the following table:

Strain No.	Streptomycin broth potency γ ml. after	
	4 days	5 days
L-118	[1]670	[1]635
X-69	860	1,480
Z-38	1,205	2,000

[1]Streptomycin yield is low for strain L-118 due to the fact that the incubation temperature is higher than the optimum value for strain L-118.

EXAMPLE 2

A 5-liter fermentor was charged with 3.2 liters of a medium containing 4% soybean meal, 0.25% sodium chloride, 0.5%

distillers solubles, 2.5% dextrose and distilled water to volume.
Approximately 0.5 P. P. M. of cobalt as cobalt nitrate was also
added to the medium, and the medium, after sterilization, was
inoculated with 5% of a vegetative culture of *S. griseus,* albus
mutant (Z-38). Another 5-liter fermentor was charged with 3.2
liters of a medium containing 3% soybean meal, 2% dextrose,
0.75% distillers solubles, 0.25% sodium chloride, distilled water
to volume and 0.5 P. P. M. of cobalt as cobalt nitrate. This
medium, after sterilization, was inoculated with 5% of a
vegetative culture of *S. griseus* Dulaney L-118. The two
inoculated mediums were incubated at 27° C. with mechanical
agitation and aeration until maximum streptomycin production
was obtained in each. At the end of the fermentation, the
following yields of streptomycin and vitamin B$_{12}$ were obtained:

| Inoculum | Streptomycin | | B$_{12}$ color isolated based on 1000 gal. volume, mg. |
	Maximum mcg./ml.	Time of maximum hours	
L-118................................	1,140	[2]113	950
Z-38	[1]1,830	[2]118	0

[1]The streptomycin yield with strain Z-38 would have been somewhat higher if the optimum temperature of 28.5° C. had been employed in this experiment.
[2]In 15,000 gal. fermentors the time for maximum streptomycin production is 80 hours for L-118 and 110 hours for Z-38.

EXAMPLE 3

An inoculum was prepared by propagating spores of *S.
griseus,* albus mutant (Dulaney Z-38) in a sterile medium
containing 1% dextrose, 1% enzymatic digest of casein, 0.6%
meat extract and distilled water to volume, under aerated and
agitated conditions at 27° C. for 36 to 48 hours until good
growth was obtained. One ml. portions of the resulting broth
were used to inoculate each of a number of flasks containing
the medium above-described, and these cultures were allowed
to grow at 27° C. under aerated and agitated conditions for 20
to 24 hours until approximately 5-7 mg./ml. of vegetative
growth (dry weight) was obtained and 3-4 mg./ml. of sugar

remained. The contents of the flasks were then pooled to provide inoculum for actual streptomycin production.

A number of 5-liter fermentors were charged with 3200 ml. portions of sterile medium containing 3.5% of soybean meal, 2.75% dextrose, 0.5% distillers dried solubles, 0.25% sodium chloride, 0.4 cc. per 100 cc. of soybean oil and distilled water to volume, which is the preferred medium for Z-38. Each of the fermentors was inoculated with 150 ml. of the prepared inoculum and incubated at a temperature of 28.5° C. for 112 to 118 hours with mechanical agitation, imparted by an impeller bearing rotary shaft, and aeration under varying conditions. The streptomycin yields obtained with the different conditions of agitation and aeration are tabulated below, the streptomycin yield in each instance being an average of values obtained in 3 or more separate fermentors.

Effect of power on streptomycin production with Z-38

Total Power Absorbed, HP/gal. of aerated fermenting medium	Superficial Air Velocity, ft./hr.	Streptomycin Yield, mcg./ml. of fermented broths
.0024	72	938.5
.0049	36	1,656
.0072	36	1,726
.0091	36	1,919
.0096	36	2,051
.0230	36	2,077

In comparison, a similar experiment was performed in the same 5-liter fermentors in which vegetative inoculum of culture L-118 was used for fermentation at 27° C. of the following sterile preferred medium: soybean meal 3.0%; dextrose 2.0%, distillers dried solubles 0.75% and sodium chloride 0.25%. Results obtained were as follows:

Total Power Absorbed, HP/gal. of aerated fermenting medium	Superficial Air Velocity ft./hr.	Streptomycin Yield, mcg./ml. of fermented broths
0.0012	36	350
0.0015	36	780
0.0024	36	1,000
0.0920	36	980

As will be noted from the above tables, increasing the horsepower absorbed per gallon of fermenting medium of 0.0024 with culture L-118 does not increase streptomycin yield above 1000 mcg./ml. of fermented broth, whereas, with the Z-38 culture such increase in horse power absorbed up to 0.009 horse power per gallon of fermented medium raises the streptomycin yield up to 2000 mcg./ml. of fermenting broth.

EXAMPLE 4

The procedure of preparing an inoculum of *S. griseus,* albus mutant (Dulaney Z-38) and using this inoculum to inoculate 3200 ml. portions of medium in a number of 5-liter fermentors was repeated as described in Example 3. The inoculated mediums were then incubated at the different temperatures as indicated with constant mechanical agitation and aeration until maximum streptomycin production was obtained. Agitation in each instance was supplied by means of 2 turbo impellers revolving at such a rate that the total power absorbed was 0.0091 HP./gal. The streptomycin yield at the different incubation temperatures employed is indicated in the following tabulation, the values in each instance representing an average of two or more separate fermentations.

Temperature, °C.	Streptomycin Yield, mcg./ml.	Time for Maximum Streptomycin Yield, hours
25	1,180	118
27	2,041	118
29	2,194	104
31	414	72

Additional experiments have indicated that the optimum incubation temperature with the organism *S. griseus,* albus mutant (Dulaney Z-38) is 28.5° C.

EXAMPLE 5

The procedure of preparing an inoculum of *S. griseus* albus mutant (Dulaney Z-38) and using this inoculum to inoculate 3200 ml. portions of medium in a number of 5-liter fermentors was repeated as described in Example 3. The mediums prior to inoculation were each adjusted to predetermined pH values by addition of caustic, i. e., sodium hydroxide solution, either before sterilization or after sterilization as indicated. The inoculated mediums were then incubated at a temperature of 28.5° C. with constant mechanical agitation and aeration until maximum streptomycin production was obtained, the agitation in each instance being provided by two turbo impellers revolving at such a rate that the total power absorbed was 0.0091 HP./gal. The fermentations were conducted as two separate experiments, one to determine the comparative effectiveness of pH adjustment before and after sterilization, and the other to determine the optimum pH adjustment before sterilization. The streptomycin yields obtained in these experiments are tabulated below, the values in each instance representing an average of two or more separate fermentations.

Effect of pH adjustment before and after sterilization

Initial pH of Sample	pH Adjustment	Time for Maximum Streptomycin Yield, hours	Streptomycin Yield, mcg./ml
6.6	7.0 before sterilization	112	1,769
6.8	7.0 after sterilization	118	1,555
6.4	6.5 after sterilization	104	1,378
6.2	6.0 after sterilization	104	1,266

Effect of different pH adjustment before and after sterilization

Initial pH of Sample	pH adjustment ml. 30% NaOH added before sterilization	Time for Maximum Streptomycin Yield, hours	Streptomycin Yield mcg./ml.
6.65	3.2 (pH approx. 7.0)	112	1,913
6.33	2.2 (pH less than 7.0)	118	1,859
6.5	4.2 (pH greater than 7.0)	96	1,799

From the foregoing tabulations, it is evident that better results are obtained when the pH is adjusted prior to sterilization and that optimum results are obtained when the pH is adjusted to approximately 7.0 prior to sterilization.

From the foregoing examples, it is evident that the following criteria, in addition to the typing test results already set forth, can be used to distinguish *S. griseus,* albus mutant (Dulaney Z-38) from its parent *S. griseus* Dulaney L-118:

	Z-38	L-118
Optimum fermentation temperature	28.5°	27° C.
Fermentation time for maximum streptomycin production.	110 hours	80 hours.
Optimum pH of medium:		
Before sterilization	7.0.......	6.2-6.4.
After sterilization	6.5-6.6 ...	6.1-6.3.
Minimum agitation power absorption for maximum streptomycin production HP/gal. fermenting aerated medium.	0.0009 ...	0.0024.
Approximate maximum yield of streptomycin—mcg. of streptomycin per ml. of fermented broth.	2,000	1,000.
Vitamin B¹² potency of fermented broth	none	950 mg. per 1,000 gal. of broth.

Various changes and modifications may be made in carrying out the present invention without departing from the spirit and scope thereof. Insofar as these changes and modifications are within the purview of the annexed claim, they are to be considered as part of my invention.

I claim:

The process for the production of streptomycin that comprises fermenting an aqueous nutrient medium under submerged aerated and mechanically agitated conditions by means of the herein described organism *Streptomyces griseus,* albus mutant (Dulaney Z-38).

EUGENE L. DULANEY.

REFERENCES CITED

The following references are of record in the file of this patent:

UNITED STATES PATENTS

Number	Name	Date
2,445,748	Demerec	July 27, 1948

Figure 85. Example of a U. S. patent covering an antibiotic-producing microorganism.

Chapter Six _____

RECOMBINATION IN BACTERIA

_____

FUNDAMENTALS OF RECOMBINATION

THE PROGRESS OF EVOLUTION would not be possible if genetic material were immutable and not subject to changes and rearrangements. Nature has provided organisms with two means by which they are afforded new combinations of genetic information. One of the means is mutations, the subject of the previous two chapters. The second means is known as *recombination,* by which organisms exchange genetic information among themselves to create new combinations of genes. In higher plants and animals recombination is the result of sexual unions, usually between members of the same species, in which the progeny resulting from the union exhibit combinations of genes not possessed by either of the parents. Among the microorganisms, recombination can occur through sexual processes, through mechanisms referred to as parasexual, or through events that defy classification. The bacteria have demonstrated three methods for carrying out recombination, one being closely similar to sexual processes in higher forms, the other two not similar. In this chapter we will cover recombination in the bacteria; in a later chapter, recombination in the viruses will be discussed. We will start off, however, with a simplified discussion of some fundamental biological principles.

The first published observation of recombination in modern times was made by the British geneticists Bateson and Punnett in 1906. In following the transmission of phenotypes through succeeding generations of garden peas, they observed devia-

189

tions from expected Mendelian ratios with regards to flower color and pollen appearance. These traits did not follow Mendel's Law of Independent Assortment, but appeared for the most part to remain together through successive generations, or were *coupled,* in the terminology of the day. That is, when plants with purple flowers and long pollen granules were crossed with plants possessing red flowers and round pollen granules, approximately 95 percent of the resulting progeny would exhibit the original parental combinations. The remaining 5 percent, however, expressed recombinant phenotypes: purple flowers and round pollen, or red flowers and long pollen. Bateson and Punnett named this phenomenon *partial coupling,* but admitted lack of insight as to its mechanism.

By 1913 experimental evidence accumulated by Morgan, Sturtevant and others at Columbia University in New York led to the establishment of a model to explain partial coupling. It was shown that genes are arranged on the chromosomes in a linear fashion, as Sutton had hypothesized in 1902. By cytological observation, the chromosomes of a given eucaryotic cell can be arranged into morphologically identical pairs, the members of each pair being known as *homologs.* Each pair of homologs, however, are distinguishable from one another by various morphological characteristics (Fig. 86).

Thus, it would appear that each chromosome pair carries its unique assemblage of genetic information in duplicate. By following the movements of chromosomes in living cells over several generations, distinct stages are observed during the division of the chromosomes and their distribution to the daughter cells. This process, known as *mitosis,* is diagrammed in Figure 87. Note that at some time prior to the stage known as anaphase, at which time chromosomes are seen to separate, the homologous pairs must replicate in order for each daughter cell to retain duplicate copies of each chromosome. A similar process, called *meiosis,* happens in the formation of gametes, the fundamental differences being that in a final division cycle, chromosome replication does not occur, resulting in the daughter cells receiving but one chromosome of each pair (Fig. 88). Cells having duplicate sets of chromosomes are known as

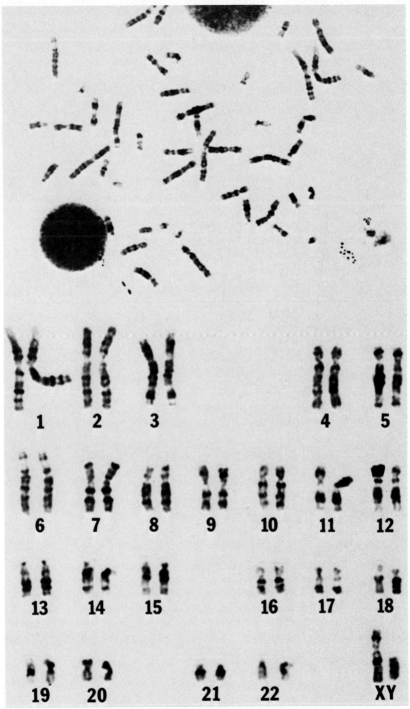

Figure 86. Human caryotype. The chromosomes of a human leucocyte are stained and photographed (upper figure). Each of the chromosome pairs are then identified according to specific morphology and arranged in a specific order (lower figure). In this manner, abnormalities are quickly spotted. (Photo courtesy Raymond Teplitz, M.D.)

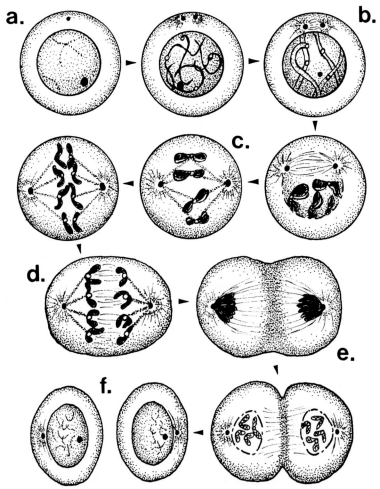

Figure 87. Mitosis. During the early stages of mitosis (prophase, a through c), the chromosomes are undergoing replication. At the stage known as anaphase (d), chromosome pairs separate and are drawn to opposite poles of the dividing cell, and at telophase (e), a nuclear membrane forms around the chromosomes. Division of the cell is then completed (f).

being *diploid,* whereas cells that have formed as a result of meiosis or for other reasons have but one member of each chromosome pair are referred to as being *haploid.* Finally, the fertilization of an ovum by a male gamete brings together the

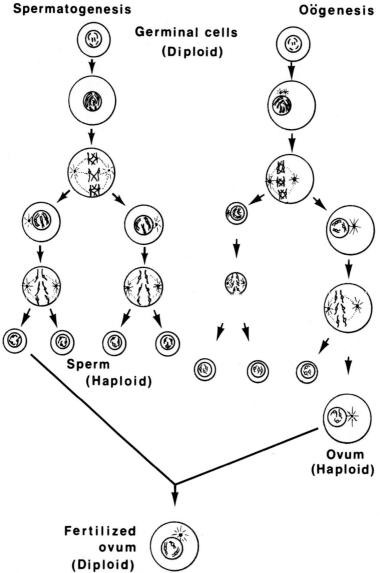

Spermatogenesis   **Oögenesis**

Germinal cells
(Diploid)

Sperm
(Haploid)

Ovum
(Haploid)

Fertilized
ovum
(Diploid)

Figure 88. Meiosis. Schematic representation of the manufacture of male and female gametes. The formation of sperm and ova begins with the division of diploid germinal cells (top) through two cycles, the last of which being a reduction division. In reduction division, the chromosomes do not divide, resulting in daughter cells that are haploid (containing but one member of each chromosome pair). Note that whereas in spermatogenesis, four gamete cells result from the one germinal cell, in oögenesis only a single ovum results, the remaining smaller daughter cells, known as polar bodies, being nonfunctional. Finally, fertilization of the ovum restores the diploid state.

genomes of each parent and restores the diploid condition in the zygote.

Any two or more genes, each occurring on a different chromosome, would demonstrate Mendelian independent assortment relative to one another. But if they happen to appear on the same chromosome, independent assortment theoretically would not be possible, but would result in the observation first called coupling, now referred to as *linkage*. However, as Bateson and Punnett and others had observed, linkage was not complete, for a fraction of the progeny exhibited a recombining of the parental traits. The mechanism of recombination was first approached by T. H. Morgan when he suggested that chromosomes can exchange information by *crossing-over*. As an example, a cell's chromosome pair $\frac{A \quad B}{a \quad b}$ containing two gene pairs, the members of which are nonidentical, may exchange genetic information in a manner that results in the new genotypic arrangement $\frac{A \quad b}{a \quad B}$. The cell still will exhibit the same phenotype AB but on division its daughter cells would receive the chromosomal complements $\frac{A \quad b}{A \quad b}$ or $\frac{a \quad B}{a \quad B}$ and express the recombinant phenotypes Ab or aB.

Two fundamental models have been proposed over the years to explain the physical basis for crossing-over. One model, generally referred to as *breakage and reunion,* is basically what Morgan had suggested, that homologous chromosomes physically break, and the broken ends are exchanged and reunited. Many observers found this model difficult to accept, for chromosome breakage usually had been considered an abnormal phenomenon associated with mutations and possible death of the cell. The possibility that two homologous chromosomes could break at exactly the same point and reunite without incurring some damage was considered too remote. One suggestion was that the chromosomes do not have to break at *exactly* the same point, and in fact it would be better if they did not. As Figure 89 depicts, if the chromosomes were to break at different points, the reunion of the broken ends is facilitated by a mechanism in which the loose ends use complementary strands as a means of aligning the rejoining pieces. Repair enzymes could then excise the loose strands and fill in gaps that might occur as a result of the operation.

The alternate mechanism for recombination involves no physical exchange of genetic material but merely an informational exchange. This mechanism was first presented by Belling in 1931, and was developed further by Lederberg in 1955, who coined the name for it: *copy-choice*. In Figure 90 we see the replication of chromosomes, where at some point in the replication cycle the replication mechanism skips from one chromosome to the other in much the same way your eye might skip to the next line while reading this sentence. The new strand thus formed will have copied part of the genetic information from one chromosome, and part of that from the other chromosome. The copy-choice model appealed to many, for it did not involve the difficult, and errorless maneuvers required of the chromosomes in breakage and reunion.

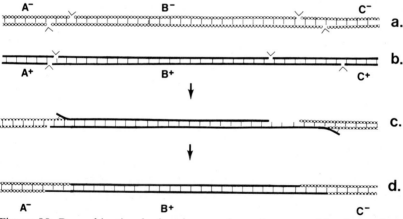

Figure 89. Recombination by breakage and reunion at nonidentical points. Homologous chromosomes (a. and b.) with genotypes A⁻B⁻C⁻ and A⁺B⁺C⁺, respectively, undergo recombination by breaking at points indicated by carets. Reunion occurs through the pairing of homologous base sequences (c.), and finally repair enzymes fill gaps and snip off excess nucleotides (d.).

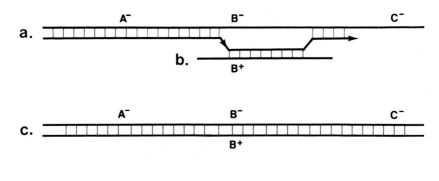

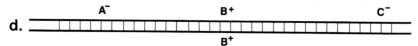

Figure 90. Recombination through copy choice. During the replication of a chromosome (a.) a fragment of an homologous chromosome, but with differing genotype (b.) may be used as the replication template. The information thus copied becomes part of the replicating chromosome (c.), which at this stage is heterozygous with respect to the B marker. An additional round of replication results in the homozygous, recombinant chromosome shown in (d.).

A number of experiments and observations have been carried out to test one or the other of the two models for crossing-over, but the most compelling evidence seems to point to the breakage and reunion model.

Of the evidence that supports breakage and reunion, that which is most often cited and is most appropriate for this volume, resulted from an experiment devised by Meselson and Weigle in 1961. This piece of research involves recombination in bacteriophage lambda. If one simultaneously infects *E. coli* host cells with two strains of lambda, each differing from the other by two characteristics, some of the progeny phage that result from the infection will exhibit recombinant phenotypes. Meselson and Weigle grew quantities of one of the strains (say with phenotype AB) in host cells in the presence of nutrients containing carbon-13 and nitrogen-15, heavy isotopes of these elements. The DNA of the phage particles thus possessed a greater density than normal lambda DNA and banded in a more dense region of a CsCl density gradient (see Chapter Two). The ^{13}C-^{15}N strain was then allowed to infect a bacterial host culture simultaneously with the other strain (phenotype ab) of normal density in the presence of normal density nutrients. After one infectious cycle was completed, the released phages were harvested and tested for density and phenotype. The results of this experiment are depicted in Figure 91.

Phage particles exhibiting the AB (parental) phenotype produced three bands in CsCl: a dense band corresponding to the density of the original double stranded DNA containing ^{13}C and ^{15}N; a medium density band corresponding to DNA that is half heavy and half normal, the result of semiconservative replication of the parental DNA; and a light band, corresponding to daughter DNA molecules produced in the presence of the normal nutrients. Phage that displayed the other parental phenotype ab exhibited a single DNA band of normal density. The bacteriophage that showed recombinant phenotypes (Ab or aB) also exhibited *three* density bands, one heavy and one medium density, but neither of which exactly corresponded to the densities of the parent AB bands, and one normal (light) density band. The question raised is how did the recombinant phage that showed heavy DNA acquire the greater density? Density cannot be copied; it must be physically inherited from the heavy parent. It must be concluded that the recombinant DNA was formed by the reassociation of fragments from the parental DNA.

As a further confirmation for the presence of the breakage and reunion mechanism in recombination, one experiment involved the crossing of two parent phages in normal nutrient media, both of which contained heavy DNA. All recombinants formed possessed DNA that corresponded in density to nearly totally heavy DNA. If the copy-choice model were operating, all recombinant DNA would exhibit an intermediate density. The small light contribution that was observed in the recombinant DNA density may have revealed some replication going on, perhaps associated with the repair of the gaps as was depicted in Figure 89.

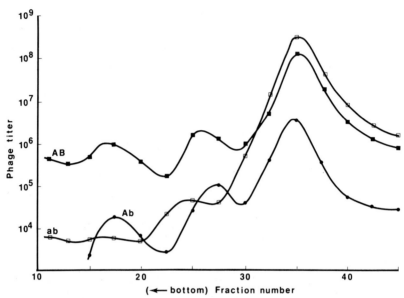

Figure 91. Results of the Meselson and Weigle experiment. Two strains of lambda bacteriophage, displaying phenotypes AB and ab, simultaneously infect a bacterial culture in normal nutrient medium. Strain AB had been propagated in a medium containing ^{15}N and ^{13}C nutrients and thus exhibited greater density. Strain ab was of normal density. The progeny phages resulting from the infection were harvested and subjected to density gradient centrifugation. Each fraction was then assayed for phenotype. The parental phage AB exhibited a density peak near the bottom of the tube, reflecting homogenously labelled particles that had not participated in the infection. Another peak of AB phages appears at an intermediate density, reflecting phage particles that had undergone one round of semiconservative DNA replication in the presence of normal density nutrients. The major peak of AB viral particles constituted the progeny virions synthesized from normal precursors. Strain ab, having been propagated in normal medium, showed a single major peak in the normal density region. Looking now at phages exhibiting a recombinant phenotype (Ab), it can be seen that a proportion of them demonstrated peaks in the higher density region, suggesting that they are composed of part heavy and part normal DNA. The conclusion reached is that the recombinant phages inherited fragments of heavy DNA from their parental virions by the process of breakage and reunion.

The breakage and reunion model demonstrated in bacteriophage lambda by Meselson and Weigle appears in its broad outline to apply to recombination in bacteria and in eucaryotic microorganisms as well. In fungi there is evidence that recombination is accompanied by extensive DNA excision and repair, again supporting that portion of the model involving uneven breakage and the filling of gaps (Fig. 89). It also appears that there are definite points on the eucaryotic chromosome where recombination is initiated. The greater complexity of their chromosomes has prevented any clear-cut picture of the mechanism of recombination in higher plants and animals, but it is presumed that it follows the same general course of breakage and reunion. It should be noted again, however, that the copy-choice model has not been eliminated entirely, and that there may be occurrences of it in nature that are yet to be uncovered.

TRANSFORMATION

Brief mention was made previously (Chapter Two) of the discovery of transformation in *Streptococcus pneumoniae* by Griffith in 1928, and of the subsequent demonstration by Avery and coworkers in 1944 that transformation was mediated by free bacterial deoxyribonucleic acid. Operationally, transformation involves the extraction of DNA from a donor strain of bacteria, and the mixing of the DNA, with or without further purification, with recipient bacteria that differ by one or more phenotypic traits from the donor. The recipient bacteria take up the DNA and later they are tested for expression of the donor traits. As a general rule, donor and recipient strains in transformation are of the same species, although transformation between different species within the same genus (*interspecific transformation*) and between members of different genera (*intergeneric transformation*) has been reported. Such heterospecific gene transfer is less efficient than that between members of the same species by factors of 10^{-1} to 10^{-7}. It has been proposed that the relative efficiencies of transformation between bacterial strains are a measure of their taxonomic relatedness. This aspect will be covered in a later section.

Since Griffith's original discovery, transformation has been shown to occur in some dozen genera of bacteria, including *Hemophilus, Streptococcus, Bacillus, Rhizobium, Neisseria, Acinetobacter,* and *Escherichia.* At least one or two genera are added to the list each year.

One of the most significant advances in the study of transformation was the discovery by Spizizen in 1961 of transformation in *Bacillus subtilis.* Up to that point in time, the only species of bacteria that demonstrated transformation were those with complex nutritional requirements, such as *Hemophilus* and the pneumococcus. These constraints limited transformation experiments to but a few markers. *B. subtilis,* in contrast, has essentially the same simple nutritional requirements as *E. coli,* making it now possible to utilize dozens of nutritional markers, as well as a bonus, those markers involved with sporulation.

More recently, transformation has been reported to occur in *E. coli,* first using bacteriophage DNA and resulting in phage production and cell lysis. This phenomenon is called *transfection* and has been carried out in other genera as well. It will be covered in greater detail in a later section. In *E. coli,* transfection or the transformation of bacterial markers appear to require special procedures. For example, recipient bacteria must be treated with *helper phages* or $CaCl_2$ before the DNA is taken up by the cells. Furthermore, success is achieved only when recipient *E. coli* cells lack the ability to form certain enzymes known to degrade linear DNA molecules, but are still capable of supporting recombination.

Competence

If a bacterial strain that is known to undergo transformation is grown in liquid medium, and aliquots are removed at intervals and exposed to transforming DNA, it will be found that early in the life of the culture few cells exhibit transformation (Fig. 92). As the culture matures, a greater proportion of the cells undergo transformation until a peak is observed, after which the proportion drops off rapidly. In *Streptococcus* the peak occurs early in the log phase, while in the bacilli, pneumococcus and *Hemophilus,* it occurs late in the log phase. The cells in the culture appear to experience a fleeting capability to take up the transforming DNA, a characteristic known as *competence.* Competence, a general phenomenon prevalent among most bacterial species known to undergo transformation, is genetically controlled, for it is possible to isolate mutants that have lost their ability to develop competence. Competence no doubt reflects some temporary physiological state of the cells, but its exact basis is not clear. We

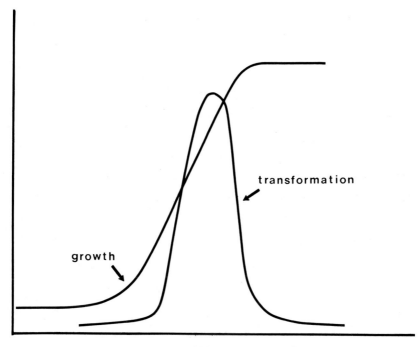

Time

Figure 92. Appearance of competence. The development of competence in a bacterial culture occurs during a short interval of the growth cycle. In this example, competence is at its peak late in the log phase.

can, however, describe some of the characteristics of competent cells.

The interval during which time a cell remains competent varies considerably depending on the species. In the pneumococcus this interval is but a few minutes in duration, while cells of *B. subtilis* appear to maintain the competent state for as much as four hours. The fraction of bacteria exhibiting competence at the time a culture is demonstrating maximal transformation also varies greatly. Nearly 100 percent of the cells in a pneumococcal culture eventually attain competence, whereas in *B. subtilis* fewer than 20 percent of the cells ever become competent. This anomaly may reflect our lack of knowledge of the conditions necessary to bring about maximal competence in the bacilli. Although competence is a phenomenon observed in all transformation systems, optimal conditions necessary to induce cells to competence are as diverse as the genera that exhibit it. For example, a relatively rich growth medium is necessary to bring about competence in the pneumococcus and in *Hemophilus.* In contrast, a rich medium inhibits the establishment of competence in the bacilli. The latter requires agitation or aeration to induce competence, the former bacteria do not.

There is evidence that the establishment and maintenance of competence requires protein synthesis. This can be demonstrated by treating recipient bacteria before or during time of peak competence with drugs known to specifically inhibit protein synthesis, or RNA synthesis. In either case, the level of competence is curtailed drastically. Some specific proteins associated with the competent state have been isolated from streptococci, *S. pneumoniae* and *B. subtilis,* and are referred to as *competence factors,* or CF's. CF's appear to be heat labile polypeptides that when added to noncompetent cultures of bacteria, induce competence in them. CF's are physically and serologically distinguishable and are active only with homologous organisms or those that are taxonomically related to the source of the CF. The target of CF activity is not completely known, but it appears to be the so-called receptor sites that are found on the surface of competent cells. The pneumococcus possesses as many as eighty such sites, *B. subtilis,* about fifty or less, while *Hemophilus influenzae* does not appear to have more than four. As their name implies, receptor sites are thought to participate in the initial binding of transforming DNA to competent cells prior to its uptake.

Spontaneous or induced mutants lacking CF or receptor site formation have been isolated from cultures of streptococci. Strains unable to make CF can still become competent if treated with CF from competent streptococci. Mutants lacking receptor sites cannot be made competent by any treatment.

Other physiological characteristics of competent bacteria that may have a bearing on the establishment of competence are as follows:

1. *Nucleases*—Competence may occur during an interval in the life of a culture when production of extracellular deoxyribonucleases is at an ebb, making it more likely for transforming DNA to approach the cell and be taken up. This notion is supported by the requirement in transformation in *E. coli* that recipient cells lack certain nucleases.

2. *Cell Wall or Membrane Integrity*—Competent *B. subtilis* cells appear to be more susceptible to osmotic shock, and possess cell walls with altered biochemical characteristics. Further, bacteria mildly treated with lysozyme are more susceptible to transformation, all suggesting that weakened or damaged cell envelopes are more permeable to transforming DNA. Transformation in *E. coli* offers support to this view in that the addition of $CaCl_2$ increases the permeability of this organism's cell membrane towards DNA.

3. *Spores and Mesosomes*—In *B. subtilis* many mutations that block spore formation also alter the establishment of competence, suggesting an association between the two activities, but this has not been substantiated. A similar association has been observed in the bacilli between competence and the presence of membranous structures known as *mesosomes*. Electron microscopic examination of cultures reveals that the competent cells exhibit a considerably greater number of mesosomes. Actual uptake of DNA by competent cells may be mediated by these

structures, and in fact they may be part of the receptor site described above.

4. *Chromosome Structure*—DNA extracted from competent *B. subtilis* cells exhibits as much as 5 percent single strandedness, as opposed to near 0 percent in noncompetent cells. While it is difficult to relate competence (defined as the ability to take up DNA) with the physical state of the recipient chromosome, it is clear that the single strandedness may increase the probability of the donor DNA becoming integrated into the host DNA.

The Stages of Transformation

Different bacterial species differ somewhat in specific details concerning transformation. However, some general statements can be made regarding the stages of this process (Fig. 93).

1. *Binding of DNA*—The first stage in transformation is the binding of a DNA molecule to the surface of a competent recipient bacterial cell. This stage, lasting about a minute, appears to consist of two major steps. Initially the binding is reversible, being sensitive to high ion concentrations and to the action of exonucleases. During this period, recipient cell endonucleolytic enzymes reduce the donor DNA to smaller double-stranded fragments of molecular weight of the order 10^7. These fragments are then quickly converted to single-stranded molecules through the action of further recipient cell nuclease activity. The binding stage is now irreversible, becoming insensitive to DNase in the medium, but it does become sensitive to cyanide and azide. The binding stage is not species specific, in that DNA from any source, including mammalian DNA for example, will compete with homologous DNA for binding.

2. *Penetration*—In a few minutes or less following the binding stage, the DNA, now single-stranded, begins to appear within the recipient cell. Its molecular weight is now about 2 to 5×10^6. Penetration is also species nonspecific. Within minutes, initial steps of the final stage, integration, can be detected.

3. *Integration*—In order for the donor DNA to be expressed phenotypically, it must be integrated into the genome of the

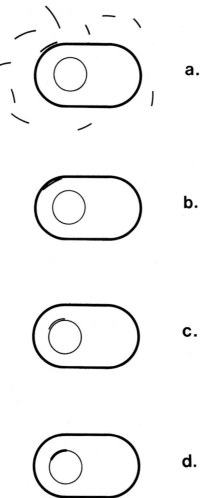

Figure 93. The stages of transformation. (a.) Donor DNA reversibly binds to recipient cell. (b.) Irreversible binding and penetration of DNA into recipient cell. At this point the donor DNA is converted to single-strandedness. (c.) Positioning of donor fragment to homologous region of recipient DNA prior to integration of donor fragment into recipient genome (d.).

recipient cell. The first step in this process is the positioning of the single-stranded donor DNA fragment along the recipient chromosome in the genetic region corresponding to the donor fragment. The association of the donor fragment with the recipient DNA, therefore, must involve a recognition mechanism. It is presumed that the association step operates on conventional Watson-Crick pairing of homologous DNA regions and that it occurs at some special time when chromosome strands are separated, such as during RNA transcription or DNA replication. Through breakage and reunion all or part of the donor fragment then undergoes recombination with the recipient DNA. The time that integration is finally achieved varies greatly from system to stystem; it may take anywhere from ten minutes to several generations for its completion.

The sizes of the transforming DNA fragments integrated into the host chromosome average about 1 to 5 x 10^6 daltons in *B. subtilis, Hemophilus,* and *Streptococcus pneumonaei.* Such fragments are roughly equivalent to about 0.1 percent of the donor genome.

Heterospecific Transformation

As we have seen, competent bacteria are capable of taking up DNA of unrelated species, but integration and phenotypic expression is limited to genetic material from the same or closely related species.

Heterospecific integration of genetic information bears great significance in areas of human health, where ultimately it may be possible to counteract a metabolic disease by replacing an errant gene with one that operates satisfactorily.

Data from transformation experiments between *B. subtilis* and other species of *Bacillus* indicate that in spite of the similar base compositions of their DNA's, there is wide variation in the degree of transformation that occurs between the different species. The extent of *in vitro* hybridization between the different DNA's has also been determined, and here we see some correlation between transformation and the ability of the donor DNA to hybridize with the recipient DNA. Since hybridization appears to be a phenomenon founded on specific, homologous base-pairing, it can be concluded that the integration of donor DNA into the recipient genome in transformation also depends on such a base-pairing mechanism, where identical or near identical base sequences are more likely to lead to the matching of corresponding genetic regions and the ultimate integration of donor DNA.

Genetic Mapping

When transforming DNA is extracted from donor cells, it generally is reduced to small fragments by enzymatic action and the shearing action of pipetting, centrifugation, and other manipulations. Whereas the molecular weight of intact bacterial DNA is of the order of 1 to 3 x 10^9 daltons, the average molecular weight of DNA molecules extracted by the method of Marmur, for example, is around 2 x 10^7 daltons. In other words, each transforming DNA molecule in a transformation experiment may carry on the average only about 1 percent of the donor genome. It appears that the DNA is broken down further by enzymatic action of the recipient cell to the point that on the average fragments representing only about 0.1 percent of the donor genome are ultimately integrated into the recipient chromosome. We mentioned briefly the possibility of carrying out the transformation of two or more markers simultaneously. For example, DNA from an ind$^+$ his$^+$ donor bacterium can be mixed with a suspension of competent recipient cells that are ind$^-$ his$^-$ and certain fractions of the recipients will be transformed to ind$^+$ his$^+$. How can this double transformation come about? The recipient cells exhibiting the two donor traits could have received the ind$^+$ and his$^+$ markers either on two separate pieces of DNA, or if the markers were close enough together on the donor chromosome, the recipient may have received them on a single piece of DNA. Put another way, it is possible to estimate the distance between two markers on a chromosome by observing the frequency of double transformation. As you can see, the closer two markers are, the greater is the probability that they will be transmitted together.

Since we can estimate the average length of the transforming DNA pieces, markers that regularly are transformed together must reside at distances less than that average length. How is it determined whether a doubly-transformed cell received the markers on a single piece of DNA, or on two separate pieces taken up simultaneously by the recipient bacterium? Suppose we observe that among 1000 ind⁻ his⁻ bacteria treated with DNA from an ind⁺ his⁺ donor, ten are singly transformed to either ind⁺ or his⁺ whereas one exhibits the double transformed phenotype ind⁺ his⁺. Thus, we find the frequency of single transformations to be 10/1000 or 10^{-2}, whereas our observed double transformation frequency is 1/1000 or 10^{-3}. If the recipient bacteria had received the two markers on separate DNA pieces, representing two separate integration events, we would expect the frequency of those events to be the product of the separate frequencies: $(10^{-2}) (10^{-2}) = 10^{-4}$. Since our observed frequency of double transformation (10^{-3}) is greater than the frequency we would expect for separate events (10^{-4}), we come to the conclusion that the markers entered the recipient cells on the same fragment of DNA; that is, they were *linked,* and their distance apart can be estimated as discussed above.

However, one thing we failed to take into consideration in our preceding example was the degree of competency of the recipient culture. In our example, if only 10 percent of the 1000 cells were competent and capable of participating in transformation, our observed frequencies for single transformation would be more correctly 10^{-1}, and for double transformations, 10^{-2}. Now when we compare the observed frequency for doubles with the expected frequency for the same with unlinked markers, we find that they are identical (10^{-2}). From these calculations we would then assume the two markers were taken up by the recipient cells on separate pieces of DNA, that they were not linked, and that they are located on the chromosome at a distance greater than the average length of a fragment of transforming DNA.

In addition to the necessity of considering the proportion of competent cells among the recipients when determining linkage by transformation, one must be aware that the concentration of donor DNA also has a bearing on frequencies of transformation. At high concentrations of DNA, there is no difference in transformation frequencies between closely linked markers and those more distantly located on the donor chromosome. However, as DNA concentrations are reduced, the numbers of transformed cells begin to drop off proportionately. The slope of the curve (transformants vs. DNA concentration) for double transformations of closely linked markers is identical to that for single transformations, but is dissimilar when the markers are unlinked. Therefore, transformation experiments must be carried out in that range of donor DNA concentrations in which the two possibilities can be distinguished. (For a further discussion of linkage in transformation see Goodgal's 1961 paper.)

While genetic mapping by standard transformation methods spans but short intervals of the bacterial chromosome, a technique involving transformation devised by Yoshikawa and Sueoka makes it possible to map the entire chromosome. Their method is based on the assumption that DNA replicates in a sequential fashion, and in the DNA extracted from a randomly dividing culture of donor bacteria, there will be a certain number of DNA fragments containing specific genes. For a given gene, the number of fragments containing that gene will be dependent upon the distance it is from the origin of DNA replication; the closer it is, the more copies of it will be found in the DNA (Fig. 94). Thus, when using this DNA in a transformation experiment, one merely has to count the number of recipient cells transformed for each gene in question, and the relative proportions will indicate their positions on the chromosome relative to the origin. There is, however, one consideration that has to be included in the calculations. Not every gene is transformed with the same efficiency. It appears that some genes are more readily integrated into the recipient genome than others. In Table XXIII we have shown the relative frequencies with which

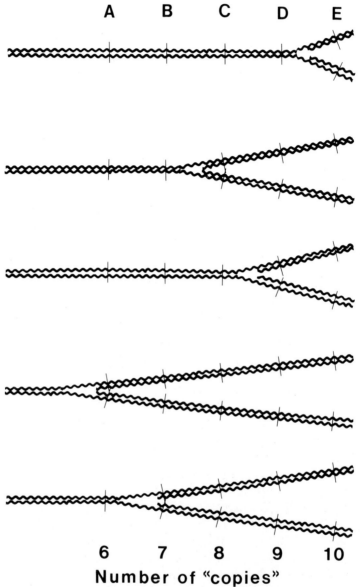

Figure 94. Genetic mapping by the method of Yoshikawa and Sueoka. In an actively growing culture of bacteria, a random sampling of the chromosomes of the cells would reveal replication in various stages of completion. In this diagrammatic representation of such a sampling, genes A, B, C, D, and E are shown, together with the number of copies of each gene. Replication is proceeding from right to left with the origin at the far right. Note that the number of copies of each gene is indirectly proportional to the distance the gene is from the origin. If these DNA fragments were used in a transformation experiment, the number of cells transformed with each of the genes would be an indication of the relative position each gene occupies on the chromosome. (From Yoshikawa, H, and Sueoka, N.: *Proc Natl Acad Sci USA*, *49:559*, 1963)

various markers are transformed using DNA extracted from a resting culture. In a resting culture, the chromosomes of all cells are at the same point of replication, and thus each marker is present in the same number. The data derived from the resting culture are then used to correct the results from the randomly growing culture experiment.

TABLE XXIII
RESULTS OF THE YOSHIKAWA-SUEOKA EXPERIMENT*

Marker	Relative frequency of transformation[†]
Adenine	1.92
Threonine	1.67
Histidine	1.28
Leucine	1.24
Isoleucine	1.04
Methionine	1.00

*From Yoshikawa, H., and Sueoka, N.: *Proc Natl Acad Sci USA, 49:*559, 1963.

[†]Transforming DNA was extracted from a wild strain of *Bacillus subtilis* during exponential growth. The DNA was mixed with various competent auxotrophic mutants and the frequency of transformation for each of the markers relative to that for the methionine marker was calculated. Those frequencies were then divided by frequencies for the same markers found by using DNA extracted from a resting culture of the wild donor. Those ratios are shown here and reflect the positions of the markers on the bacterial chromosome relative to the methionine marker.

Transfection

If DNA is extracted from bacteriophage particles and mixed with appropriate bacteria, some of the cells will eventually lyse and release mature phage particles. *Transfection,* as this phenomenon is called, was first demonstrated in *E. coli* by Fraser, Mahler, Shug, and Thomas in 1957, and in *B. subtilis* by Romig in 1962. Transfection and transformation have many parallel characteristics. Both involve purified DNA, and both generally require recipient cells to be in a state of competence, although the time of maximal competence for transfection may not necessarily coincide with that for transformation. The efficiency of transfection is considerably less than that of transformation for several reasons. The number of phage genome molecules that must be mixed with recipient cells to produce one infected bacterium ranges from about 5×10^3 to over 10^6.

Under certain conditions, transfection can be enhanced by special treatment of the recipient bacterial cells. Simultaneous infection by a defective *helper phage* has been used in *E. coli* and *B. subtilis* to improve transfection efficiency. The role of the helper phage is not clear; it is not capable of lysing host cells alone. It may play a part in altering the host cell's permeability to DNA, or it may undergo recombination with damaged, transfecting DNA that leads to the formation of infectious DNA and the ultimate lysis of the cell. The latter idea is supported by the observation that bacteria infected with helper phage will exhibit transfection by phage DNA that has been fragmented into pieces that may be as small as about one fifth of the normal. These fragments will not support transfection in the absence of the helper phage.

Transfection in *E. coli* by phages lambda, φX174, and T1 DNA has been possible by promoting spheroplast (protoplast) formation prior to addition of DNA. Transfection with phage M13 RNA has also been demonstrated in *E. coli* spheroplasts. Other treatments, such as protamine sulfate and ultraviolet radiation, will likewise increase the level of transfection.

Practical Considerations

Laboratory manipulations for demonstrating transformation are relatively simple in established systems. DNA can be extracted from the donor strain by any of a number of techniques, such as the buffered-phenol procedure of Saito and Miura. Attaining competent recipient cultures usually requires growing the bacteria in specific media under carefully controlled conditions for several generations prior to exposure to the transforming DNA. The best markers to be transferred are those for which selection is convenient, such as drug resistance or prototrophy. For example, DNA from a prototrophic strain of *B. subtilis* is mixed with a suspension of competent *B. subtilis* auxotrophs. Recipient bacteria are then spread on minimal agar plates and the degree of transformation is reflected in the number of colonies formed compared with a control culture not exposed to the DNA.

To attempt the demonstration of transformation in species in which this phenomenon has not been shown before presents severe problems. Transformation has been readily demonstrated in both Gram-negative and Gram-positive bacteria, but Gram-negative varieties generally require special treatment with helper phage or $CaCl_2$ before becoming competent. Notable exceptions to this rule appear among members of the genera *Neisseria* and *Acinetobacter.* To discover the conditions necessary to bring about the competent state in a given species is especially troublesome, for as was implied in an earlier section, such conditions are largely unpredictable. A recommendation that can be made is to try those conditions that work with related species, keeping open the possibility of the unexpected. Transformation in *Acinetobacter,* for example, was accidentally discovered when various auxotrophic strains were cross-streaked on minimal medium plates to test for syntrophism.

Finally, great variability has been observed among different strains of the same species with respect to their capacities to undergo transformation. If a given bacterial strain fails to demonstrate transformation, it is highly recommended that other strains be tried.

Despite the fact that transfection has been shown in far fewer species than has transformation, it would appear that transfection would be the easier to demonstrate. Manipulations to bring about competence for transfection essentially are identical to those for transformation, keeping in mind, however, that their peaks may not coincide. The use of helper phages was previously discussed. Helper phages generally are defective, conditional mutants of the wild type from which the transfecting DNA is extracted. The helper phages cannot complete the infection under the conditions of the transfection experiment. Soon after exposure of recipient cells to the transfecting DNA, the recipient cells are plated onto a lawn of indicator bacteria that are capable of detecting the presence of the resulting phage particles. This technique is covered in Chapter Seven.

In addition to their usefullness in genetic mapping, transformation and transfection are excellent tools to assay the effects of various physical and chemical treatments on the biological activity of purified DNA. As an example, transfecting DNA can be exposed to extreme ionic strengths, hydrostatic pressures, or gamma radiation, and then applied to an appropriate transfection assay system. Note can be made of the effect of the treatment on biological activity, or in addition, on its mutagenicity.

CONJUGATION

Single-celled organisms, specifically protozoa, have been widely known to undergo recombination through a process known as conjugation. Conjugation is defined as sexual reproduction through direct cell contact. Cells such as those of *Paramecium* are seen to unite pair-wise at which time the cells' micronuclei have divided meiotically. Three of the resulting four daughter nuclei disintegrate, the fourth divides once mitotically, and one of the daughter nuclei is exchanged between the conjugating pairs. Thus, each cell receives a nucleus from its partner, and the fusion of the donor nucleus with the nucleus of the recipient cell completes the mutual exchange of genetic information. The pairs separate and the *exconjugants* may be tested for possession of recombinant genotypes. Conjugation among protozoa was observed in the 1690's by Leeuwenhoek, and its genetic implications were explained in the early decades of this century, principally by Sonneborn and his collaborators.

Many attempts were made to demonstrate conjugation in bacteria, even prior to the birth of bacterial genetics in the early 1940's. The most frequently cited report of negative results in this regard is that of Sherman and Wing in 1937. In 1946 Lederberg and Tatum successfully displayed conjugation in a strain of *E. coli*. The success of Lederberg and Tatum where others failed was based on good experimental design. These workers chose to employ auxotrophic mutants which, when forming rare prototrophic recombinants through conjugation, were easily detected on minimal media. You recall that Tatum had used auxotrophs of *Neurospora* in the establishment of the one gene-one enzyme hypothesis with Beadle in the early 1940's. He extended this experience to the isolation of numerous auxotrophs of *E. coli*. To eliminate the possibility of back-mutants being mistaken for prototrophic exconjugants, double and triple mutants were utilized. Thus, whereas the probability of a single auxotroph (his$^-$) reverting to prototrophy may be 10^{-7}, the probability of a double auxotroph (his$^-$thr$^-$) reverting to prototrophy is the product of the individual back mutation rates, 10^{-14}, obviously unlikely odds. We must not overlook the ever present element of chance, for of the thousands of bacterial strains available to Lederberg and Tatum in the Stanford University culture collection, they chose one of the few strains of *E. coli* that is capable of conjugation, K-12. Following the initial discovery of bacterial conjugation, 2000 additional strains of *E. coli* were screened by Lederberg, out of which but fifty exhibited conjugation.

Bacterial conjugation as carried out by Lederberg and Tatum simply involved mixing together two different auxotrophs, possessing phenotypes such as $A^-B^-C^+D^+$ and $A^+B^+C^-D^-$, and plating the mixture onto minimal agar on which neither of the mutants could grow (Fig. 95). Members of colonies that did appear were subcultured, had their phenotypes tested, and proved to exhibit stable recombinant phenotypes $A^+B^+C^+D^+$. The possibility that transformation may have played a part in the observed recombinations was dismissed by these workers by carrying out the following control experiments. These experiments involved the addition of cell-free extracts of one of the partners to the growth medium of the other, with no recombination resulting, and the addition of DNase to the mixture of the two mutants, which had no effect on the appearance of recombinant prototrophs. Furthermore, Lederberg observed that the occurrence of double and triple recombinants was too frequent to be due to transformation.

Some years later, Davis devised a more positive technique to dismiss the possibility of transformation and to prove the necessity of direct cell contact for bacterial conjugation. Davis constructed a U-shaped tube the arms of which were separated

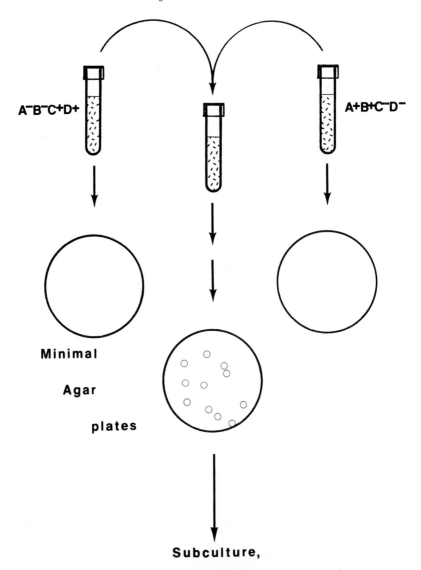

Subculture,

Confirm recombinant genotype (A⁺B⁺C⁺D⁺)

Figure 95. Demonstration of conjugation in bacteria. Two dissimilar double auxotrophic mutants, with genotypes $A^-B^-C^+D^+$ and $A^+B^+C^-D^-$ are incubated together, following which the suspension is spread onto minimal agar plates. Neither of the mutants can grow on the minimal medium, but significant numbers of colonies appear when inoculated with the mixture. The colonies are subcultured and tested for recombinant genotype.

by a filter of sintered glass. The pores of the glass were sufficient for solutes (including DNA) to circulate freely between the arms, but not for the passage of intact bacterial cells (Fig. 96). To increase circulation of solutes, alternating mild pressure and vacuum were applied to one arm. When each of the auxotrophic mutants was placed in one of the arms of the U-tube, no recombination was observed. When the experiment was repeated with an identical tube, but without the dividing glass filter, recombinant types were readily recovered.

Conjugation appears to be a widespread phenomenon among many families of bacteria, such as *Enterobacteriaceae (E. coli, Salmonella, Shigella), Vibrionaceae (Vibrio), Pseudomonadaceae (Pseudomonas),* and *Rhizobiaceae (Rhizobium).* Interspecies and

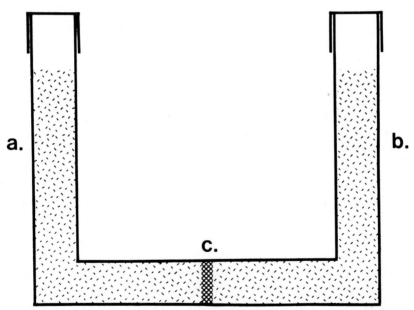

Figure 96. The Davis U-tube. Conjugation in bacteria is confirmed by growing each parent strain in one of the side arms of the U-tube. A sintered glass filter (c.) prevents direct cell contact, but allows the flow of liquid and macromolecules from one arm to the other. With the filter in place, no recombination by conjugation is observed, whereas when it is removed, conjugation is possible.

intergeneric recombination through conjugation has been demonstrated repeatedly.

The Fertility Factor

It had been assumed by Lederberg and others that conjugation in bacteria was a homothallic affair in which the role of donor and recipient was readily exchanged. However, the British bacteriologist William Hayes carried out a series of experiments in which it appeared that the transfer of genetic information in bacterial conjugation was heterothallic and unidirectional. In one of Hayes' *E. coli* strains, in addition to the usually auxotrophic markers, the marker for streptomycin resistance was added. The genotypes of the participants of a typical cross were:

(1) $A^-B^-C^+D^+Str^r \times A^+B^+C^-D^-Str^s$

in which Str^r referred to streptomycin resistance and Str^s to sensitivity towards the antibiotic. In another cross, the markers for streptomycin resistance and sensitivity were exchanged:

(2) $A^-B^-C^+D^+Str^s \times A^+B^+C^-D^-Str^r$

To Hayes' surprise, cross (1) resulted in a normal number of recombinants whereas cross (2) yielded no recombinant colonies on minimal agar containing streptomycin. The streptomycin appeared to prevent the appearance of recombinants in cross (2) but not in cross (1). Hayes saw this as an indication of a unidirectional transfer of genetic information, in which the strain possessing the $A^+B^+C^-D^-$ genotype acted as the donor, while the alternate strain was the recipient. Thus, streptomycin would have no effect on the recipient in cross (1) and recombinants would appear. In cross (2) the recipient is streptomycin-sensitive and succumbs to the antibiotic. (Apparently the Str^r marker of the donor was not transferred to the recipient under the conditions of this experiment.) Hayes set out to explain what determines whether a cell acts as a donor or a recipient in conjugation.

It eventually became clear that the capacity to act as a donor cell in bacterial conjugation depended upon the intracellular presence of a virus-like particle named the F-factor (F=

fertility), or sex factor. The infectiousness of the F-factor was manifested in Hayes' observation in 1953 that bacteria apparently lacking the fertility factor, i.e. F⁻, rapidly acquired it when incubated in liquid media with an F⁺ strain. Conjugation was definitely involved in transmission of the F-factor, for cell-free extracts or lysates would not convert F⁻ cells to F⁺. These observations were confirmed independently by Cavalli, Lederberg, and Lederberg in 1953.

A more striking observation made by Hayes was that contact of F⁻ cells with one particular strain of fertile bacteria resulted in little conversion to fertility when they were incubated together, but this same donor strain exhibited a thousand times higher capacity to transfer certain bacterial markers on conjugation. This strain was named Hfr (High Frequency of Recombination). In contrast, those F⁺ strains that readily transferred the fertility factor to F⁻ cells were poor in their ability to transfer bacterial markers. Thus, there appeared to be an interrelationship between the F-factor and the bacterial genome, one that was eventually explained in the following manner: The F-factor is an extrachromosomal, independently replicating piece of genetic material that falls under the general appellation of plasmid (see Chapter Seven). In F⁺ bacteria the plasmid is found in the cytoplasm, but in the Hfr bacteria on the other hand, the F-factor has become integrated into the bacterial chromosome. Thus, when an F⁺ cell conjugates with an F⁻ partner, usually only the F-factor is transferred. In the Hfr cell, the association of the F plasmid with the donor chromosome induces the transfer of the chromosome, but most of the F plasmid remains behind in the donor cell. The lack of transfer of the F-factor in a Hfr x F⁻ cross appears to be due to the manner of breakage of the chromosome in preparation for transfer. As depicted in Figure 97, the donor chromosome breaks at a point that is actually within the F plasmid itself. The donor chromosome is transferred to the F⁻ cell as a single strand, in linear fashion; but most of the F-factor is at the trailing end of the chromosome. Thus, the fertility factor is not transferred to the F⁻ cell *in toto* unless the entire chromosome of the donor is transferred, which happens only infrequently.

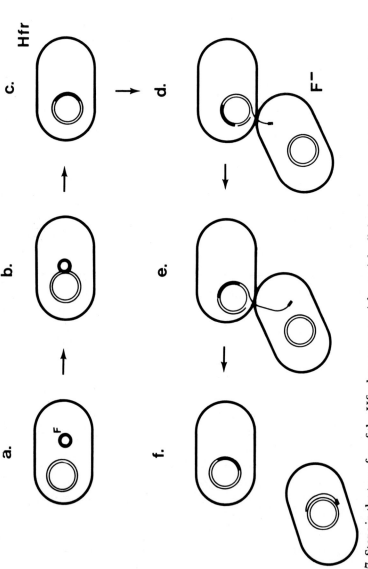

Figure 97. Steps in the transfer of the Hfr chromosome. A bacterial cell that has received the F-factor (a.) may be converted to an Hfr cell through the integration of the fertility plasmid into the cell chromosome (b.). Contact with an F⁻ cell (d.) initiates the linear transfer of one strand of the Hfr chromosome, starting at the integrated F-factor. Replication of donor and recipient DNA may occur during transfer (only donor DNA replication is shown). At some random time following initiation of the transfer, conjugation may be interrupted, at which time the F⁻ cell departs with all, or more probably part of the Hfr chromosome, and about one third of the F plasmid. Unless the F⁻ cell receives the entire Hfr genome, it does not gain the entire F-factor and thus remains F⁻.

In spite of the fact that there appears to be no bacterial genetic material associated with the F plasmid in the F$^+$ state, some bacterial markers are observed to be transferred in an F$^+$ x F$^-$ cross. The conversion from F$^+$ to Hfr is spontaneous, and it is likely that in any given culture of F$^+$ bacteria, one will find a few Hfr bacteria. It is these Hfr cells that may account in part for the low level of recombination seen in the F$^+$ x F$^-$ cross. There is also evidence that the F plasmid itself may indirectly bring about the mobilization and transfer of some bacterial genes without having been associated with the bacterial chromosome.

There is yet a third conditon that is observed in fertile bacteria. In the Hfr state, the fertility factor is integrated into the donor chromosome. There are times when the F-factor breaks out of its integrated state and returns to its former cytoplasmic station. In doing so, the act of excision frequently results in an aberrant detachment in which one or more bacterial markers remain attached to the fertility factor. This condition, known as F' (F-prime), was discovered by Adelberg and Burns in 1960 and is characterized by transfer of the fertility factor in crosses with F$^-$ cells (Fig. 98), as well as bacterial genes.

It has subsequently been shown that the point of integration on the bacterial chromosome of the fertility factor is characteristic of the particular strain of donor cell. Thus, depending on the donor strain from which it originated, F' factors may carry any of the markers found on the chromosome.

A summary of the characteristics of crosses involving the three states of fertile bacteria is shown in Table XXIV.

Figure 98. Formation of the F' state. The F-factor of an Hfr cell reverses the integration process, but in doing so, retains some bacterial DNA. Conjugation between F' and F$^-$ cells thus results in the transfer of a few bacterial genes as well as the F plasmid.

TABLE XXIV
COMPARISON OF TYPES OF CONJUGATION IN *ESCHERICHIA COLI*

Type	Relative frequency of recombination	Number of bacterial markers usually transferred	Usual sex of recipient following conjugation
F⁺ x F⁻	low	0 to few	F⁺
F′ x F⁻	moderate	few	F⁺
Hfr x F⁻	high	few to many (may be entire chromosome)	F⁻

Steps in Bacterial Conjugation

Experiments to determine the molecular events that occur during conjugal transfer of plasmid or bacterial genomes have been exceedingly difficult to design. We can, however, briefly outline the steps that characterize bacterial conjugation.

The ability to act as a donor cell in conjugation rests on the presence of the F-factor, which itself is acquired through conjugation. The F-factor appears to be a circular particle of double-stranded DNA with a molecular weight of 4.5×10^7 daltons. This is sufficient genetic material to code for about forty genes. It is, as all plasmids have proved to be, dispensable to the cell for normal cell functions. Part of the F genome is responsibile for the independent replication and maintenance of the plasmid. Certain genes are involved in the expression of sexuality in the bacterial cell and others in the sequence of steps that occur during conjugation. One of the early functions of the F-factor is to induce the bacterial cell to produce surface appendages known as F-pili. The F-pili, which are antigenically distinct from pili normally produced, appear to act as specific attachment organs in the formation of conjugal pairs with F⁻ cells. This view is supported by electron microscope observations and by experiments in which the pili of Hfr bacteria have been removed or chemically altered, the result of which was the elimination of the ability to conjugate. Conjugation is restored at the same rate that the pili reappear. Furthermore, bacteriophages have been discovered that specifically attach to F-pili and their presence in a mating mixture severely reduces the level of conjugation. Presumably the phage particles block the functioning of the pili.

When fertile donor bacteria possessing normal F-pili are mixed with F^- cells in sufficient numbers, mating pairs are quickly formed. The attractive forces maintaining conjugating bacteria together are moderately strong; pairs usually will remain attached during mild agitation, such as that accompanying the gentle preparation of dilutions. Pairs do detach spontaneously, however, which gives rise to the probability function regarding the transfer of a given gene relative to its distance from the leading end of the chromosome. Violent mechanical agitation, such as that produced by "Vortex" mixers, Waring Blenders® or similar laboratory apparatus leads to near complete separation of mating pairs. In addition to being specific organs of attachment, the F-pili also have been implicated as conduits for the transfer of DNA to the recipient bacteria. The structure of pili make this hypothesis attractive: They are hollow, proteinaceous tubes, with an axial opening of about 25 Å diameter.

Once contact is made with a recipient cell, the mobilization of the F^+ or F' factor, or the donor chromosome, is initiated. Mobilization involves the preparation of the appropriate DNA molecule for length-wise transfer into the recipient bacterium, events which are under the control of the F-factor genome.

In the case of an Hfr x F^- cross, the donor chromosome enters the recipient cell as a single strand. Thus, one of the mobilization events involves the breaking of one strand of the donor's DNA and its unwinding. It has been extremely difficult to dissect the events that occur during the actual transfer of DNA. It appears that DNA replication in the donor cell is necessary for the initiation of chromosome transfer, but replication in the recipient cell is required for the maintenance of the transfer process. It has been suggested that the replication of the incoming strand in the recipient cell acts as a pulling force to effect the transfer. Integration of donor genetic information into the recipient cell then seemingly occurs by the breakage and reunion recombination mechanism discussed earlier.

Mapping

As previously mentioned, Lederberg observed that two or more markers were transferred with remarkably high frequency in bacterial conjugation, leading him to conclude that quite a large fragment of bacterial chromosome was involved. Lederberg extended the mutant genotypes of his experimental organisms to as many as five markers (A B C D E) but selected for only certain ones on the plating media. For example, he might carry out the following cross: $A^-B^-C^+D^+E^+$ X $A^+B^+C^-D^-E^-$ and select for those recombinants expressing the phenotype $A^+B^+C^+D^+$. By including the metabolite required by the E^- mutants in the screening medium, both E^+ and E^- cells would grow on the minimal medium so long as they possessed the genotype $A^+B^+C^+D^+$. The resulting colonies are then tested for the presence of the E^+ marker among their members. This technique is known as screening for *unselected* markers, and makes it possible to record with greater precision the frequency with which certain genes are transferred together.

Within a year following the initial discovery of conjugation in bacteria, Lederberg was able to start the mapping of the *E. coli* chromosome. Lederberg's method was based on the aforementioned probability function associated with a given gene's distance from the origin: The closer a gene is to the leading end of the chromosome, the more likely is its transfer. However, the technique of interrupted mating devised by Wollman and Jacob in 1955 offered a considerable improvement in bacterial chromosome mapping.

Chromosomal Mapping by Interrupted Mating

The transfer of the bacterial chromosome in a cross between Hfr and F^- strains occurs in a highly ordered, linear fashion. It is therefore possible to estimate the locations of particular genes on the bacterial chromosome by determining the time at which they enter the F^- cell. A strain of Hfr bacteria with genotype $A^+B^+C^+D^+E^+$ is incubated with an F^- strain that is

A⁻B⁻C⁻D⁻E⁻, for example, and at regular intervals (five minutes), aliquots of the mixture are removed, subjected to violent agitation to separate the mating pairs, and plated on appropriate media to determine the phenotypes of the recipient exconjugants. The media are specially prepared to

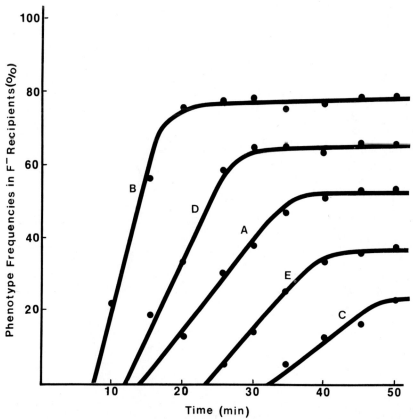

Figure 99. Genetic mapping by interrupted mating. A prototrophic Hfr strain is mixed with an F⁻ strain that is multiauxotrophic (for example, A⁻B⁻C⁻D⁻E⁻). At regular intervals, samples are taken from the conjugation mixture and plated onto various supplemented minimal media to determine the presence of prototrophic markers in the F⁻ cells. The curves that result indicate the order in which the markers enter the F⁻ cells, and by extrapolating the curves to the baseline, the time at which the markers enter is also determined.

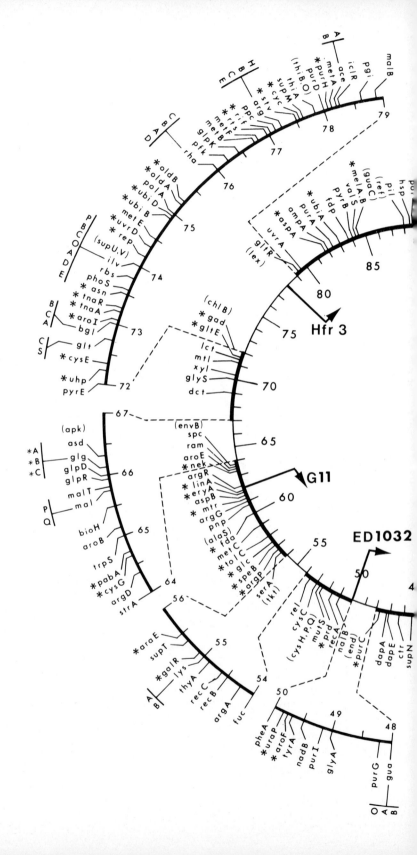

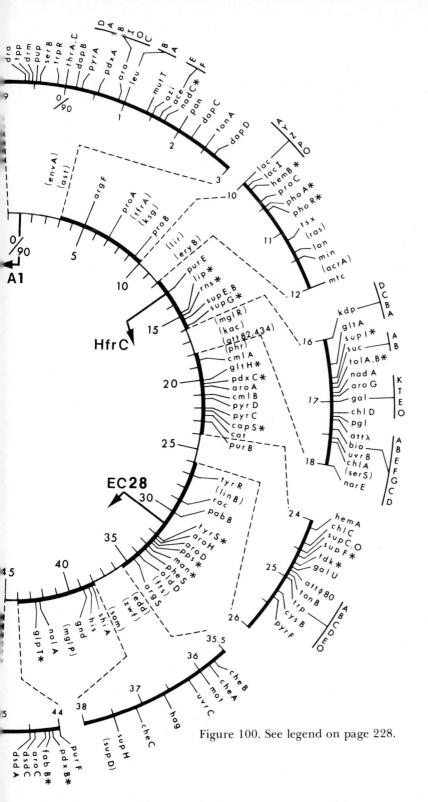

Figure 100. See legend on page 228.

Figure 100. Circular map of *E. coli* genome. The locations of some 450 genetic markers are shown on this circular map. The inner circle is divided into 90-minute segments, the total period necessary for the entire chromosome to be transferred by an Hfr strain under standard conditions. In many cases, segments had to be expanded to accommodate large numbers of loci. Expanded segments are shown in the outer circle. Asterisks and parentheses indicate degrees of uncertainty as to the exact location of the marker, parentheses indicating least certainty. Arrows adjacent to the symbols for certain operons, such as *his* and *bio,* indicate the direction of mRNA transcription.

Within the inner circle are also shown the locations of integration of various F episomes, together with the direction of genome transfer on conjugation.

The identity of some key markers are as follows:

Marker	Location (minutes)	Function
ace	2,80	acetate requirement
ampA	83	resistance to penicillin
ara	1,55	arabinose utilization
arg	various	arginine synthesis
aro	various	aromatic amino acid synthesis
attλ	17	lambda prophage attachment site
att	various	attachment sites for prophages shown
azi	2	azide resistance
bio	18,66	biotin synthesis
cml	19,21	chloramphenical resistance
cys	various	cysteine synthesis
dna	various	DNA synthesis or repair
gal	17, 27, 55	galactose utilization
hag	37	H flagellar antigen
his	39	histidine synthesis
hsm ⎫ hsr ⎬ hss ⎭	89	host controlled modification functions
ilv	48, 75	isoleucine-valine synthesis
lac	9	lactose utilization
met	various	methionine synthesis
mut	2,52,83	mutator genes
phr	17	photoreactivation
pil	88	pili formation
pol	2,4,76	DNA polymerases III, I & II
pur	various	purine synthesis
pyr	various	pyrimidine synthesis
rec	51,54,73	ultraviolet sensitivity and recombination
str	7,64	streptomycin resistance

Marker	Location (minutes)	Function
sup	various	sites of suppressor mutations
thi	78,79	thiamine synthesis
*ton*A	3	resistance to phages T1 and T5
tsx	10	resistance to phage T6
uvr	various	repair of UV damage to DNA
xyl	70	xylose utilization

(From Taylor, A.L., and Trotter, C.D.: *Bacteriol Rev, 36:*504, 1972, where a complete catalog of markers can be found.)

eliminate the Hfr strain, usually by incorporating an antibiotic and arranging for the Hfr to be sensitive, and the F⁻ to be resistant. The results of such an *interrupted mating experiment* are shown in Figure 99. It can be seen that the markers appear to enter the F⁻ cells in a sequential fashion, the time of entrance being dependent upon their location on the bacterial chromosome. Thus, a chromosome map may be constructed by charting the time interval in which given genes enter the F⁻ cell. By shortening the intervals to one or two minutes, greater precision in locating the positions of specific genes may be achieved, although a point is reached where asynchrony of matings begins to result in ambiguous results.

We stated earlier that the point of integration of the fertility factor in Hfr strains differs from strain to strain. This fact is supported by observing the sequence in which various markers are transferred by different Hfr strains. The data shown in Table XXIV enables one to see that this hypothesis is supported. This observation was made by Jacob and Wollman in 1957 and was the first indication that the bacterial chromosome may be a closed, circular molecule.

Figure 100 depicts the genetic map of the *E. coli* chromosome, showing the locations of over 450 genes, plus the origin sites of some Hfr strains. This map, published by Taylor in 1972, is divided into ninety one-minute segments, the minutes referring to the transfer time scale under standard conditions into the F⁻ cell as determined by the interrupted mating technique.

Assuming the *E. coli* chromosome to consist of 5 x 10^6 nucleotide pairs, the transfer of the single-stranded Hfr DNA operates at the rate of nearly 1000 nucleotides per second. As pointed out above, interrupted mating does not have sufficient resolution to distinguish the positions of genes that are much less than a couple of minutes apart; a span that theoretically represents the genetic material of 200 genes. Transduction, a technique to be covered in the next chapter, is used to determine the finer genetic structure of the chromosome.

A similar genetic map has been prepared by Sanderson for *Salmonella typhimurium,* showing the locations of some 350 genes. Interrupted mating was also the basic technique utilized here, with transduction supplying the details. A different procedure generally is applied to interrupted mating experiments with *Salmonella,* for the binding of mating pairs is exceedingly weak. To avoid wholesale, premature interruption of mating in conjugating strains of this genus, the pairs usually are spread onto solid agar media or membrane filters. Matings are interrupted by washing the cells off of the supporting surface and plating onto appropriate growth media. As in *E. coli,* there exist Hfr strains of *Salmonella* with differing points of integration of the F-factor, although the number of such strains that have been found is considerably less. The standard transfer time in *Salmonella* is 138 minutes (on solid media) compared with ninety for *E. coli* (in broth). There are some similarities between the maps of the two organisms, underlining their putative evolutionary relationship.

BACTERIAL VIRUSES AND PLASMIDS

INTRODUCTION TO THE LITTLE THINGS IN LIFE

VIRUSES ARE SMALL, infectious agents that are distinguished from other microorganisms by possessing structural simplicity exemplified by a central core nucleic acid and a protein coat. The coat, or *capsid,* is made up of structural units called *capsomeres,* which in turn consist of many individual protein molecules. Some viruses have lipid and other materials associated with them. A mature virus particle is known as a *nucleocapsid* or *virion.* Viruses are obligate intracellular para- sites, but this is not a principal distinguishing feature, for many other microorganisms such as the rickettsiae, also possess this characteristic. The manner of viral replication within the host cell is unique, however, in that components are synthesized separately and subsequently assembled into virions utilizing a combination of viral and host enzymes and cofactors.

Although not fully confirmed, all living cells probably play host to one or more viruses. Viruses are generally classified according to type of nucleic acid (DNA or RNA) and its strandedness, and according to the geometric arrangement of the capsomeres and other characteristics (see Chapter Eight). Viruses also can be arranged according to host; bacterial viruses (or bacteriophages), plant viruses, and insect viruses are examples of this kind of classification. This system is becoming less satisfactory however, in light of newer information that many viruses can move from members of one of these groups to members of another group. For instance, some plant viruses also appear to infect invertebrate hosts such as insects and nematodes.

The bacteriophages are the most extensively studied of the viruses. The ease with which they are propagated and handled is no doubt the reason for their popularity. Considerable attention will be focused on these viruses in this chapter. Bacteriophage-mediated transfer of bacterial genetic information, called *transduction,* will be covered also. A number of similarities exist between bacteriophages on the one hand, and plasmids and episomes on the other, so much so that the discussion of these extrachromosomal genetic elements is included in this chapter on the viruses. We have already introduced the fertility plasmid and its role in bacterial conjugation. Other types of plasmids to be discussed are those involved in the transfer of antibiotic resistance and the formation of bacteriocins, specific inhibitory agents produced by many bacteria.

TABLE XXV
CHARACTERISTICS OF NUCLEIC ACIDS OF SOME BACTERIOPHAGES

Bacteriophage	Nucleic acid type	molecular weight	genetic capacity
Fd	ss DNA	1.3×10^6	5
φX174	ss DNA	1.7×10^6	7
Lambda	ds DNA	3.3×10^7	65
T5	ds DNA	7.7×10^7	150
T2	ds DNA	1.2×10^8	250
MS-2	ss RNA	1×10^6	4
Qβ	ss RNA	9×10^5	4

ss = single-stranded; ds = double-stranded
Genetic capacity based on assumption that total genome codes for polypeptides of average molecular weight 35,000, requiring 750 nucleotides per polypeptide. Only one strand of ds nucleic acid is assumed to be transcribed.

Where nature has allowed the monotony of a single type of nucleic acid to occur in the viruses, she has compensated for it in the diversity of forms in which the nucleic acids appear. Nowhere is the diversity more remarkable than in the bacterial viruses, where nearly every permutation covering DNA or RNA, strandedness, circularity, redundancy, and other features is found. Table XXV illustrates this point, and Table XXVI goes into more detail regarding the nature of various phage DNA's. Among the phages T2 and T4, one finds the

TABLE XXVI
TYPES OF NUCLEIC ACIDS IN BACTERIOPHAGES

Phage	Permuted	Terminal Redundancy	Nature of bacteriophage
T2, T4	yes	duplex	virulent
T3	no	duplex	virulent
T5	no	?	virulent
T7	no	duplex	virulent
P1	yes	duplex	temperate
P22	yes	duplex	temperate
Lambda	no	exposed	temperate
P2	no	exposed	temperate
φ29	no	exposed	virulent

abcdefghijk

abcdefghijk

abcdefghijk

abcdefghijk

non-permuted

abcdefghijk

abcdefghijk

abcdefghijkabc

abcdefghijkabc

duplex redundancy

abcdefghijk

abcdefghijk

abcdefghijk

defghijkabc

exposed redundancy

defgjoklabc

defghijkabc

permuted

ghijkabcdef

ghijkabcdef

chromosomes

random

viral DNA to be *permuted.* That is, each linear phage genome within a population of phage particles may begin at a different point. The DNA of other phages, such as T5, is *nonpermuted;* each DNA molecule is identical to every other. Phage DNA molecules may contain redundant sequences amounting to 3 percent or less of the total genome. In some instances the redundant region may be *exposed,* that is, single stranded. In other cases, as in T2 and T4, the repetitious sequence is double-stranded, or *duplex.*

While fewer details are known about the nucleic acids of the animal viruses, it is clear that great diversity is shown here as well. Both types of nucleic acids are represented, as single strands and as double strands. The animal viruses will be more fully discussed in Chapter Eight.

THE BACTERIOPHAGES

Bacterial viruses, or bacteriophages, were discovered apparently independently by the English bacteriologist Twort and the French-Canadian d'Herelle in 1915 and 1917, respectively. Because of military duty, Twort was unable to pursue his discovery, but d'Herelle spent a number of years on the investigation of the bacterial viruses. He developed many of the techniques used to this day for the propagation and assay of bacteriophage particles, and in fact coined the word *bacteriophage.* The 1970 catalog of the American Type Culture Collection (ATCC) lists over seventy species of bacteria for which bacteriophages are available, and there may be many times that number yet to be discovered and catalogued. As is the case with most viruses, the host range of bacteriophages is narrow, usually confined to species, or even strains within species.

Bacteriophage particles can be classified into three morphological groups: the structurally complex tadpole forms, and the simpler spherical and filamentous forms. Examples of each are shown in Figure 101, and some of their physical characteristics are listed in Table XXVII. The tadpole form

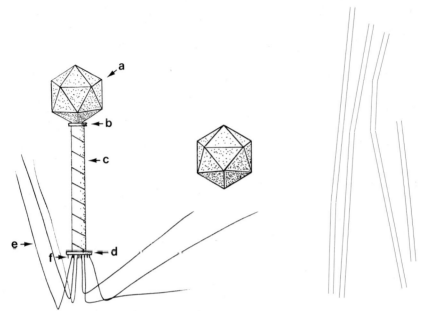

Figure 101. Fundamental bacteriophage morphologies. Left side: Typical "tadpole" morphology, with head (a.), collar (b.), tail (c.), tail plate (d.), tail fibers (e.), and tail spikes (f.).
Center: Icosahedral virion.
Right side: Morphology typical of the filamentous bacteriophages.

may consist of an icosohedral or spherical head in which the nucleic acid is found, and a tail. Much of the complexity associated with the tadpole forms is due to the tail, which plays a role in the injection of viral nucleic acid into the host cell. The simpler virus forms do not possess such an apparatus, and the mechanics of nucleic acid injection are not clear. As in all the viruses, bacteriophages possess one type of nucleic acid, either DNA or RNA.

Two alternative courses of events may occur when a bacterial cell is infected by an appropriate bacteriophage particle. The initial step in any infection process is the attachment of the phage virion to the cell surface via specific adsorption sites, followed by the passage of the phage nucleic acid into the bacterial cell. We can look at the phage nucleic acid as serving

TABLE XXVII
PHYSICAL DIMENSIONS OF SOME BACTERIOPHAGES

Bacteriophage	Morphological type	Dimensions (Å)
T2	Icosahedral head (prolated)	650 x 950
	Tail	250 x 1100
T3	Icosahedral head (isometric)	470 x 470
	Tail	100 x 150
Lambda	Icosahedral head (isometric)	540 x540
	Tail	100 x 1400
T5	Icosahedral head (isometric)	650 x 650
	Tail	100 x 1700
Fd	Filamentous	60 x 8000
φX174	Icosahedral head (isometric)	300 x 300
MS-2	Icosahedral head (isometric)	240 x 240

two functions: as a depository of genetic information from which mRNA is transcribed for the formation of phage specific protein, and as a Watson-Crick template for the replication of more nucleic acid. In the instance of the more familiar tadpole viruses like T2 and T4, the entrance of the phage nucleic acid is accomplished by injection through the phage tail. The mechanics of nucleic acid uptake during infection by the spheroidal and filamentous viruses is less clear. Within minutes following nucleic acid ingress, the decision is made as to which of two alternate courses is followed. The most familiar course is the *lytic response.* The phage nucleic acid is replicated perhaps a hundred-fold or more, and at the same time the phage genome directs the anabolic machinery of the host cell to synthesize the proteins that constitute the enzymatic and structural components necessary to assemble mature phage particles. Finally, the lysis of the cell and release of those particles is observed.

A second response, referred to as the *lysogenic response,* is seen only with certain phages. In the lysogenic response, soon after the phage nucleic acid has entered the host cell, the phage genes responsible for initiating the lytic response are repressed. In most cases the phage genome becomes integrated into the host chromosome as an episome, in a state known as a *prophage.* Following the pattern of development of other episomes, the prophage is replicated along with the host genome, so that all progeny of the originally infected bacterium will carry the prophage.

Those bacteriophages that respond only with the lytic cycle are known as *lytic,* or *virulent phages,* whereas others that are known to respond with the lysogenic cycle as well as the lytic response are called *temperate phages.* The decision as to whether a temperate phage elicits a lytic or a lysogenic response on infecting a bacterial cell depends on a number of factors to be discussed below.

Let us now take up the individual steps in the lytic and lysogenic responses in more detail.

The Lytic Response

The first step in all virus infections is the attachment of the viral particle to the outer surface of a susceptible cell. The attachment process is mediated through specific organs on both virus and host cell. In many of the tadpole phages, the attachment organs appear to be the tail fibers (see Fig. 101), while for the other morphological classes of bacteriophage, the adsorption sites presumably are located on the protein capsid. Bacterial host cells possess attachment sites that are serologically distinguishable and specific for given bacteriophages. In a few cases the sites of phage attachment are flagella or pili, an example of which is found in bacteriophage f2, which only attaches to the F-pili of fertile *E. coli* cells.

Environmental factors play a significant role in the adsorption of phage particles to host cells. The ionic nature of the suspending medium is particularly important, especially regarding the presence of mono- and divalent cations. Optimum concentrations for adsorption of T1 phage, for example, appear to be 10^{-2} M for monovalent ions, and 5×10^{-4} M for divalent ions. For the most part, the specific ion seems immaterial; Mg^{++} can be replaced by Ca^{++}, Ba^{++} or Mn^{++}, for example. However, some phages appear to require specific ionic adsorption factors. In the case of phage P1, Ca^{++} is specifically required. Certain phages require particular organic cofactors for attachment. Phage T4 requires tryptophan, the action of which appears to be the mobilization of its tail fibers.

Nucleic Acid Uptake, Transcription and Replication

One of the most remarkable aspects of the T-phages and others is the manner by which they inject their DNA into host bacteria. Once attachment is completed, the phage tail tip, with the aid of phage lysozyme, penetrates the cell wall and membrane through a contraction of the tail structure, and the DNA then is injected into the host cell through the tail. Within minutes, replication and transcription of host DNA has been interrupted, and transcription of the phage nucleic acid begins instead.

The entire phage genome does not appear in most cases to be transcribed simultaneously. A definite sequence is followed in which certain early genes are first transcribed, followed by subclasses of intermediate genes, and finally the so-called late genes. Such a sequence of events is seen in coliphage T4 infections, wherein certain proteins appear immediately, but the formation of which is turned off after about ten minutes, or others after twenty minutes. Other proteins may be made after about five minutes until twenty-five minutes, and still others are not made until twenty minutes following the initiation of the infection. Those proteins made immediately are responsible for the mobilization of host material for viral building blocks, such as nucleases that break down host DNA, and the enzymes involved in the formation of phage-specific nucleotides.

Some bacteriophages require nucleotides not normally found in bacterial cells, such as the deoxy-5′-hydroxymethylocytosine triphosphate of T2 and T4 DNA. In this case, specific enzymes for the purpose of forming these nucleotides are produced early in the infection under the control of the phage genome. Table XXVIII lists some phage-specific enzymes. Phage DNA synthesis begins about four to five minutes after infection and reaches a peak in another five minutes. Within the next few minutes, major phage structural proteins begin to appear. If host cells were to be prematurely lysed at this point (10 to 12 minutes postinfection), a few infectious virions would begin to emerge. By twenty to twenty-five minutes following phage DNA injection, assembly of mature virions is near completion,

TABLE XXVIII
SOME T4 BACTERIOPHAGE-SPECIFIC ENZYMES*

Enzyme	Function	Time of Formation[†]
Nucleases	Destroy host DNA	Early
dCMP triphosphatase	Converts host dCTP to dCMP	Early
dCMP deaminase	Converts dCMP to dUMP	Early
dCMP hydroxy-methylase	Converts dCMP to dHMP	Delayed early
Thymidylate synthetase	Converts dUMP to dTMP	Delayed early
DNA polymerase	Polymerizes nucleotides to DNA	Delayed early
Glucosyl trans-ferases	Glucosylates dHMP	Delayed early
Lysozyme	Lysis host	Late

dCMP = deoxycytidinemonophosphate; dCTP = deoxycytidinetriphosphate; dUMP = deoxyuridinemonophosphate; dHMP = deoxyhydroxymethylcytidinemonophosphate; dTMP = deoxythymidinemonophsophate.

---

*Compiled from Radding, C. M.: *Annu Rev Genet, 3:*363, 1969. Reproduced by permission of Annual Reviews, Inc.
[†]Early: Formation begins immediately on infection and continues for about 10 minutes, then ceases.
Delayed early: Formation begins 2 to 5 minutes following initiation of the infection, and continues for about 10 minutes, then ceases.
Late: Formation begins about 10 minutes following initiation of infection, and continues until host lysis.

and host cells begin to lyse as a result of the delayed formation of a phage lysozyme, releasing viral particles into the medium.

The exact mechanism behind the sequential control of bacteriophage genes is not known. In some cases evidence for both positive and negative control systems as discussed in Chapter Three has been discovered. For example, in bacteriophage T7, the product of one of its genes acts as an activator of host RNA polymerase, changing the specificity of the latter to transcribe phage DNA only. In another case, phage T5, at least some control is brought about by the two-staged injection of phage DNA. The first DNA fragment controls the early events of the infection, destruction of host DNA and stoppage of host protein synthesis. In three to five minutes, the balance of the T5 genome enters the cell, an event ordered by genes of the first stage. Soon phage DNA replication is initiated by genes of the second stage, but here sequential control at the transcriptional level as seen in phages T2 and T4 above also is in evidence.

The sequential control of bacteriophage genes appears to be a general phenomenon, occurring in DNA and RNA phages alike, and in complex and simple viruses. An example of the latter is the RNA phage R17, whose genome appears to consist of but three genes, one coding for an RNA replicase, one for the coat protein, and one so-called maturation protein, whose function is not clear. The replicase is formed early in an infection by R17, but is soon shut off. Both coat and maturation proteins are made throughout the infection, but the former is produced in exponential quantities, whereas the latter accumulates in a linear fashion. Thus, not only do we see the control of the sequence of gene activity, but also the control over the kinetics of gene product formation as well.

It should be made clear that the host cell does not become completely inactivated during a phage infection. Its energy generating system remains intact to supply ATP to drive the synthesis of phage nucleic acid and protein. The host cell also supplies the tRNA, certain cofactors and the ribosomes necessary to form bacteriophage protein. Some phages have been shown to produce at least some of their own tRNA molecules, but these appear to be expendible in that mutants lacking this ability still can carry out normal infections in some host strains.

The steps of the lytic response are diagrammed in Figure 102.

The Lysogenic Response

The attachment of temperate virions to host cells and the uptake of nucleic acid seem identical to those steps observed with lytic phages. There also appear to be no major differences between the course of events in the lytic cycle of a virulent phage and the lytic cycle of temperate phage. As a matter of fact, the early steps in the establishment of the lysogenic state are similar to those followed in a typical lytic cycle. In the temperate bacteriophage lambda (λ) for example, following injection of its DNA, those early genes involved with the lytic

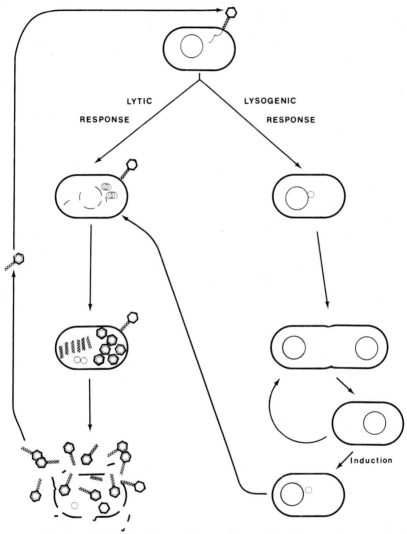

Figure 102. Lytic and lysogenic responses in bacteriophages. When a phage particle infects a host cell, the lytic response may ensue (left side of figure) where the metabolic capability of the bacterial cell is directed by the genetic capacity of the virus to the formation of a hundred or more viral particles identical to that which initiated the infection. The final outcome is the lysis of the host cell and the release of the phage particles. An alternate response is known as lysogeny in which the infecting phage genome usually becomes integrated into the host genome as an episome. Such an arrangement may last indefinitely, but on occasion spontaneous induction occurs, where the integrated phage genome (known as a prophage) is released from the host chromosome, and the lytic response is initiated.

cycle, DNA replication and recombination, and gene regulation begin to produce their respective products. In addition, one of the markers active during this time is the structural gene *int*, which is responsible for the formation of a protein that catalyzes the integration of the prophage into the host genome. Within a few minutes the products of two other genes, designated the cII and cIII genes, begin to accumulate and eventually activate gene cI. The product of the cI gene is a protein repressor that halts further progress towards the lytic response, the formation of head, tail, and lysis proteins, and at the same time, inhibits further activity of phage enzymes already formed. A repressed phage genome, the first stage in the establishment of the lysogenic state, is maintained only by the continuous, uninterrupted activity of the cI gene. The final stage in the establishment of lysogeny is the integration of the prophage into the bacterial host chromosome (Fig. 102).

Prophage Integration

Once the lytic propensities of the temperate genome have been repressed, the phage chromosome becomes a free element in the host cell. In some cases, such as in phage P1, or mutants of lambda, the prophage may remain as a plasmid, an independently replicating cytoplasmic genetic particle. For most temperate phages, however, true lysogeny cannot be established until the prophage is integrated into the host chromosome. With phages lambda and P22, the genomes first appear to form circular molecules, following which they eventually locate specific attachment sites on the host genome. The normal attachment site for the lambda prophage in *E. coli* K12 is between the galactose region and the biotin marker (Fig. 103). P22 exclusively attaches in the proline region of *Salmonella typhimurium* between markers *proA* and *proC*. In the temperate phage P2, the route to establishing the lysogenic state is basically similar to that of lambda and P22, except that the P2 prophage has more than one specific integration site on the host chromosome.

In lambda, integration of the prophage into the host chromosome occurs in the manner suggested by Campbell (Fig. 103) catalyzed by the product of the phage *int* gene, the enzyme *integrase*. As can be seen, integration requires the breakage and reunion of both host and prophage DNA. Here the prophage remains, becoming replicated along with the host DNA and being distributed to all daughter cells.

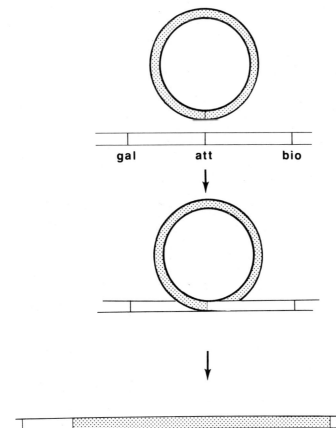

Figure 103. Integration of the lambda chromosome into the bacterial chromosome. Through a process of breakage and reunion, the circular phage genome is linearly integrated into the bacterial genome at the *att* marker, between the *gal* and *bio* genes.

Properties of Lysogenic Bacteria

Bacteria that carry a normal prophage generally possess certain features not observed in nonlysogenic strains:

1. *Immunity to Superinfection*—Lysogenic bacteria are usually immune to infection by bacteriophage identical or related to the phage that established the state of lysogeny. Thus, *Salmonella* lysogenic for P22 will fail to become infected if exposed to P22 or related phages at some subsequent time. Immunity is to be distinguished from resistance to infection, for resistance involves the inability of phage to attach to the host cell. Immune cells, on the other hand, experience phage attachment and nucleic acid injection, but no vegetative cycle will ensue. Immunity no doubt reflects the power of the prophage repressor towards invading phage genomes sensitive to it.

2. *Induction*—Lysogenic bacteria experience low probabilities of spontaneous reversion to the lytic cycle. The benign relationship between prophage and host bacterium does not necessarily last indefinitely, for the prophage may enter the lytic cycle at any time, a phenomenon known as *induction*. The observation in the 1920's that certain bacterial cultures always seemed to have phage particles associated with them, no matter what purification steps were carried out, was what led to the discovery of lysogeny. The term *lysogeny* points to this feature of lysogenic bacteria: the continual ability to exhibit lysis. In addition to induction occurring spontaneously, the rate of which may be of the order 10^{-2} to 10^{-5} per cell per generation, it may be effected artificially with certain physical and chemical agents such as ultraviolet light or mitomycin C. These agents are known to inhibit host DNA synthesis, which in turn releases a host product that, in lambda for example, inactivates the repressor and elicits the lytic cycle.

3. *Phage Conversion*—Some lysogenized strains of bacteria exhibit one or more special traits not shown by noninfected strains, a feature known as *phage* or *lysogenic conversion*. The term phage conversion is preferred, for while the phenomenon was discovered in a lysogenic system, it may also occur during a

purely lytic response. The most striking examples of phage conversion have been the formation of the exotoxin in pathogenic strains of *Corynebacterium diphtheriae,* the type C toxin in *Clostridium botulinum,* the erythrogenic toxin in streptococci, and the fibrinolysin in staphylococci. In each of these instances, the production of toxin is dependent upon the bacteria being lysogenized by particular temperate bacteriophages, or in some instances, by being infected and lysed by virulent mutants of them. Other examples of alterations due to phage conversion have been in phage typing, somatic antigen, and antibiotic resistance patterns.

A considerable amount of information has been collected on the role of phage conversion in toxogenesis in the diphtheria bacillus. It has been shown that the structural gene responsible for the toxin is located in the genome of the temperate corynebacteriophage β*tox*⁺. The toxin, a protein of molecular weight 62,000 daltons, apparently is not essential in any phage-related function involved in the establishment or completion of either the lytic or lysogenic response. Mutants of phage β (β*tox*⁻) have been isolated that carry on normal phage functions, but that fail to induce toxin formation in their host bacteria.

Pseudolysogeny

Cultures of bacteria have been known to mimic lysogenic cultures by the continued appearance of phage particles without total lysis of the culture. For example, a young culture may become infected by a small number of lytic bacteriophages, eventually resulting in the release of more phages into the medium. However, during the growth of the culture, the uninfected bacteria may have acquired a temporary, phenotypic resistance towards the phage. This may have been brought about by a phase variation as described for *Salmonella* in Chapter Three. Soon an equilibrium point is established in which there are always subpopulations of infected bacteria, of susceptible bacteria, and of resistant bacteria, as well as a constant population of free phage. Another type of pseudolysogeny involves what is called a *carrier state.* In this instance, phage are free to infect all bacteria present in the culture, but the infections may persist over several bacterial division cycles in which only one of the daughter cells of an infected parent eventually lyses. Such an arrangement leads to a constant supply of susceptible but uninfected bacteria during the life of the culture. Prolonged incubation with phage antiserum will eventually lead to a pure, virus-free culture of susceptible bacteria in the case of pseudolysogeny, but not in true lysogeny.

Zygotic Induction

Much has been learned concerning the nature of the lysogenic state through conjugation experiments with Hfr strains that were also lysogenic for λ. For example, in the following cross:

$$\text{Hfr}(\lambda^+) \times \text{F}^-(\lambda^-)$$

in which interrupted mating was carried out, F⁻ recipient cells would experience phage induction and lysis on receiving that section of the donor chromosome including the λ prophage. This reaction, known as *zygotic induction,* is due to the λ prophage being transferred from the Hfr cell in which a high level of cI repressor was maintained, to the F⁻ cell, where there was none. The transferred cI gene could not make enough repressor before the λ lytic genes were derepressed and the lytic cycle was irretrievably initiated. In the crosses:

$$\text{Hfr}(\lambda^+) \times \text{F}^-(\lambda^+)$$
$$\text{or} \quad \text{Hfr}(\lambda^-) \times \text{F}^-(\lambda^+)$$

zygotic induction is not observed because in both cases high levels of cI repressor are encountered in the F⁻ cells.

Replication of Bacteriophage Nucleic Acid

Regardless of whether a lytic response is the result of infection by a virulent phage or through the induction of a prophage, the replication of the bacteriophage nucleic acid appears to be identical.

Most of the enzymes needed to carry out phage DNA replication are produced within minutes following the initiation of the phage infection. However, actual DNA synthesis does not start until some seven or eight minutes after infection, leading to the conclusion that a trigger or signal must be activated in order for DNA synthesis to begin. While the nature of the signal is not known, it has been found that protein synthesis is necessary for its activation. In phages T4 and λ, replication procedes in a manner such that so-called *concatemers* of phage DNA are formed. That is, before the DNA is encapsulated into the mature phage head, it is in the form of large molecules equivalent in length to several viral genomes. The phage DNA is then reduced to head-full sized fragments, and condensed into packets the size and shape of which allows encapsulation into the protein viral capsid.

A few bacteriophages, notably φX174 and S13 possess single-stranded DNA in the mature virions. The nature of φX174 DNA replication in the host cell is particularly unique. The single-stranded + strand on entering the host cell, becomes the template for the formation of a complementary − strand. The resulting circular, double-stranded molecule, called the *replicative form* (RF), undergoes several rounds of semiconservative replication until a dozen or so double-stranded molecules have accumulated. The RF molecules now attach to a replication site and revert to an asymmetric mode of replication in which only + strands are produced.* The several hundred strands manufactured are then incorporated into phage heads to produce mature virions.

Certain other details of the replication of φX174 bacteriophage are of interest. Transcription takes place on the − strands of the RF molecules, meaning that viral mRNA is identical in base sequence to the parental + strands. Also, the parental strands are never found in progeny phage particles, indicating that they remain associated with their replication site

---

*A type of rolling circle replication mechanism may be involved here.

throughout the asymmetric replication mode. Finally, it appears that host enzymes are responsible for the formation of the RF molecules, but both host and virus coded enzymes are involved in the synthesis of the + strands for incorporation into mature virions.

Host Controlled Modification

Table XXIX demonstrates some instances where a certain bacteriophage may infect and lyse a host strain in column A but will generally fail to lyse a different strain of the same host in column B. It is found that the host in column B produces a restriction endonuclease that destroys the DNA of the infecting phage on its penetration of the host cytoplasm. We see here an example of immunity in which normal attachment and penetration occur, but phage development is shortly arrested.

TABLE XXIX
SOME BACTERIAL STRAINS THAT SHOW HOST CONTROLLED
MODIFICATION

Bacteriophage	*Hosts*	
	A	B
Lambda	*E. coli* C	*E. coli* K-12
Lambda	*E. coli* K-12	*E. coli* K-12 (P1)*
T1, T3, T7, P2	*Shigella* sp.	*Shigella* (P1)*
T1	*E. coli* B	*E. coli* K-12(P1)*

*Indicates host carrying P1 prophage

A remarkable aspect of such immunity through restriction is that some bacteriophage genomes appear to escape the effects of the restriction enzymes and are able to complete the lytic cycle, and the progeny of the genomes now have full capability to infect and lyse the restrictive host. If, however, these viral particles are propagated on a nonrestrictive host strain, they lose their ability to overcome the restriction immunity. Such behavior is shown schematically in Figure 104. The ability of the minority of viral genomes to survive restriction has been traced in most instances to a modification of the viral DNA such that it becomes resistant to restriction enzyme attack. The nature of the modification appears in some instances to be the methylation of adenine in the DNA in regions that are the target of the restriction enzymes. The modifications are maintained in those progeny viral genomes that issue from the restrictive host.

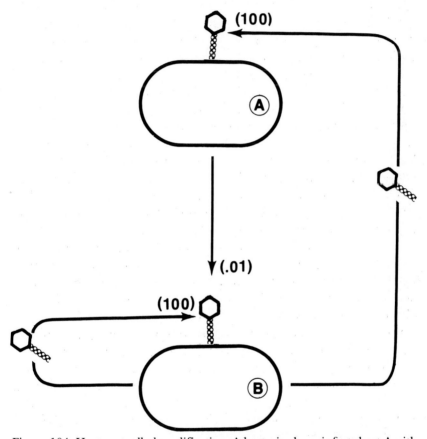

Figure 104. Host controlled modification. A bacteriophage infects host A with 100% efficiency (every phage results in a plaque). The progeny of this infection are assayed on host B, but with only .01% efficiency. Those phage issuing from host B now infect host B with 100% efficiency, as they do host A. However, once passing again through host A, the phage reacquire their low efficiency of plating on host B. Host B imparts a modification upon the phage that allows them the high plating efficiency. Passage through host A, which cannot bring about the modification, and at which time the phage cannot duplicate the modification, causes the loss of the modification and subsequent low plating efficiency on Host B.

TRANSDUCTION

During the years following the discovery of conjugation in *E. coli* by Lederberg and Tatum, other genera of bacteria were examined for the presence of this phenomenon. A Lederberg graduate student, Norton Zinder, took on a search for conjugation in the genus *Salmonella*. In one particular pair of auxotrophic strains (LT-22 as donor, and LT-2 as recipient), the degree of recombination observed appeared significant. In a routine control experiment, Zinder placed strain LT-22 in one arm of the Davis U-tube, and strain LT-2 in the other. Following several hours of incubation, the contents of each arm were tested for the presence of prototrophic recombinants. Large numbers of prototrophs were found in the arm containing strain LT-22, but none in the other arm. This unexpected outcome was carefully examined by Zinder. Conjugation was immediately eliminated by thorough leak testing of the filter that separated the arms of the U-tube. Transformation also was eliminated by showing recombination to occur in the presence of deoxyribonuclease. Zinder did find what was referred to as a *Filtrable Agent* that appeared to have originated in the LT-22 strain and was responsible for the observed recombination. In 1952 Zinder and Lederberg reported that the Filtrable Agent had all of the physical and chemical characteristics of bacteriophage particles and proposed a new mode of bacterial recombination that is bacteriophage mediated: *trandsuction*. Transduction appears to be a common phenomenon among many genera of bacteria, including *E. coli, Salmonella, Shigella, Pseudomonas, Staphylococcus, Proteus, Bacillus,* and others.

Types of Transduction

In the case where any of the several hundred genes found on the bacterial chromosome is transferred by transduction, it is referred to as *generalized transduction*. This is the type of transduction discovered by Zinder and Lederberg in *Salmonella*.

Generalized transduction involves the random inclusion of any bacterial genes into the heads of maturing phage particles during the lytic process. When the viruses that are subsequently released infect other bacterial cells, the DNA carried is injected into the cells, at which time it may undergo recombination with the recipient genome in the manner we have discussed many times before. Zinder's *Salmonella* strain LT-22 was lysogenized by bacteriophage P-22. These viruses would move freely into the arm containing bacterial strain LT-2 and infect these organisms. A few of the phage particles that were released as a result of cell lysis carried some bacterial genes with them. The pumping of the medium in the U-tube eventually carried the phages back into the side containing the LT-22 bacteria, where they infected some of the cells of this strain. A small fraction of the phages that happened to carry bacterial genes possessed the genes that were required to form recombinant prototrophic cells.

A few years later, Morse, Lederberg and Lederberg discovered that temperate phage λ limited its transducing capability to one or two specific genes: galactose or biotin. This phenomenon, known as *restricted transduction*, is due to the specificity of the integration site of the λ prophage.

There are parallels between recombination involving the F' state and restricted transduction involving lysogeny, for in both instances the integration and replication of the episomal nucleic acid is maintained at specific locations on the bacterial chromosome, and the detachment of the nucleic acid can result in one or more bacterial genes accompanying the released episome. In bacteriophage λ, the site of integration of its prophage is located between the galactose and biotin markers, which leads to the transduction of these markers only.

Transduction of the galactose (gal) marker by λ phage is conveniently demonstrated by first preparing a transducing lysate. A galactose-positive culture of *E. coli* that is sensitive to λ is infected with this phage and sufficient time is given for the establishment of lysogeny. The cells are then induced and the resulting phage particles are harvested. A heavy suspension of a gal⁻ strain of *E. coli* is then infected with a large number of the transducing phage and spread onto galactose EMB agar. Following suitable incubation, a few red gal⁺ colonies will appear against the confluent, pink gal⁻ background (Fig. 105). One gal⁺ colony results from about every 10^6 λ phages.

If one now picks each of the gal⁺ colonies and purifies it by restreaking, one observes the following: The bacteria found in about two thirds of the gal⁺ colonies appear to be unstable in their possession of the gal⁺ trait, for they produce gal⁻ daughter cells at the rate of about 10^{-3} per cell per generation. If a culture of these bacteria is induced, the resulting bacteriophage are exceptionally efficient transducers, in that a gal⁺ colony will appear for nearly every λ particle, as opposed to one out of 10^6, as in our original lysate.

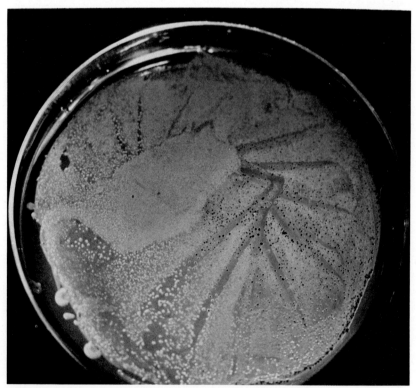

Figure 105. Gal⁺ transductants against a gal⁻ background.

The remaining one third of the gal$^+$ colonies consists of stable transductants. They appear to retain the gal$^+$ trait for indefinite generations. If these bacteria are induced, the resulting phage exhibit the same low frequency of transduction as our original lysate, namely 10^{-6}. As a result, the phage from the stable population are referred to as the LFT (Low Frequency Transducing) viruses and the phage from the unstable population as the HFT (High Frequency Transducing) viruses.

The difference between the two populations of transduced bacteria is due to the manner by which the λ prophage is associated with the host chromosome. In the unstable population, the λ prophage does not become integrated into the bacterial chromosome as we described in an earlier section, but forms a loosely-joined attachment in the form of a *super coil* (Fig. 106). Thus, the bacterial cell becomes a *heterogenote* in which its genotype can be expressed as gal$^-$/λ$-$gal$^+$. The bacterium's phenotype is gal$^+$, and it appears lysogenic in that it is immune to superinfection. During multiplication of these

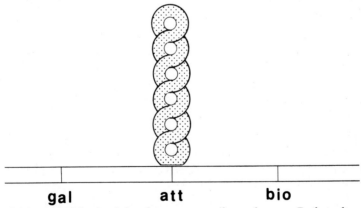

gal　　　　**att**　　　　**bio**

Figure 106. The lambda phage supercoil attachment. Rather than undergo the integration into the host chromosome (Fig. 103) the lambda genome alternatively may become loosely attached to the host genome in the form of a *supercoil.*

cells, the λ prophage is still replicated and distributed to daughter cells, but at a rate of about 10^{-3} the attachment is broken and the λ prophage with its gal⁺ marker is lost to that cell. Induction does not occur; those cells losing the λ prophage survive but no longer ferment galactose, and appear phenotypically as gal⁻, as do their progeny. If the cells carrying the λ supercoil are induced, however, the resulting phage exhibit a high frequency of transduction because every prophage still has the gal marker associated with it.

In the stable bacterial population, the λ prophage is integrated into the host chromosome in the traditional linear fashion of episomes. If these host bacteria are now induced, the harvested λ phage will exhibit the LFT trait because as the λ prophage "loops-out" prior to entering the lytic cycle, there is only the 10^{-6} probability that the prophage will take the adjacent gal marker with it.

Defective Transducing Bacteriophage λ

By including the *gal* (or the *bio*) marker in its genome, the λ prophage sacrifices a portion of its own genome, presumably because of limited volume of the λ phage head. The genes left behind are those involved with the lytic cycle, but not with the establishment of lysogeny. Thus, a gal⁺-transduced bacterium becomes lysogenized at least to the degree that it is immune to superinfection, but if it is subjected to an inducing agent, no phage particles will be produced. These λ particles are referred to as λ dg, for *d*efective, *g*alactose transducer. If λ dg particles lack lytic capabilities, how does one obtain lysates of them? Most transducing lysates of λ consist of near equal numbers of λ dg and normal λ. Thus, when preparing a lysate, unless the ratio of phage to bacteria (the *multiplicity of infection*) is much less than about 5, there is a good probability that each bacterium actually will be infected simultaneously by at least one λ particle and one λ dg particle. The normal λ phage will thus act as a *helper* virus, supplying the genetic information necessary for the lytic response of the λ dg particle as well as its own.

Abortive Transduction

What happens if a transducing particle does not become integrated into the host genome, but remains in the cytoplasm as a nonreplicating plasmid? Instances of this possibility have been observed, in which the transduced trait still is expressed phenotypically by those daughter cells that linearly inherit the particle. A striking demonstration of abortive transduction is seen in the instance of motility, where nonmotile mutants attain motility through transduction. However, the motility gene is not integrated into the recipient genome, resulting in a trail of nonmotile daughter cells leading away from a motile parent cell (Fig. 107).

Mapping by Transduction

Generalized transduction involves the transfer of any of the bacterial genes. It has been observed, as it has in transformation, that there remains the possibility of multiple gene transfer in transduction. That is to say, two or more bacterial genes may be incorporated in a single transducing bacteriophage particle. This is made possible by the fact that generalized transducing phages such as P1 may contain nothing but bacterial DNA. Thus, during an infection of the donor cell by phage P1, the probability of two or more markers being incorporated into the same phage is dependent upon their relative nearness to one another. That the phage particles cannot handle a piece of bacterial DNA any larger than what can be enclosed in their heads puts a very distinct upper limit on the distance between linked markers, as well.

Transduction has been an extremely valuable tool for determining the distances between very closely spaced bacterial genes. As an example, through interrupted mating experiments, it is determined that five genes are located in the interval zero to one minute, but their exact order or distances from one another is not known. For simplicity, let us designate the genes as A, B, C, D, and E. A donor strain of *E. coli* with

256

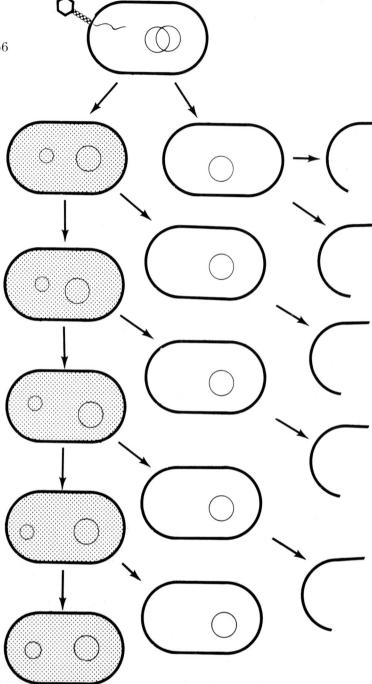

Figure 107. Abortive transduction. Diagrammatic representation of the phenomenon of abortive transduction, in which a transducing genome is passed to a recipient cell, but instead of normal integration taking place, the prophage remains in the cytoplasm and does not replicate. While the cell may express the phenotype of the transducing particle (shading), only one of its progency receives the transduced gene, the trait is inherited linearly and is soon lost through dilution.

genotype $A^+B^-C^+D^+E^+$ is infected with the generalized transducing phage P1 and the viruses are harvested. A recipient strain is prepared with genotype $A^-B^+C^-D^-E^-$ and exposed to the transducing lysate. Recipient cells are then spread onto various differential media to determine the extent of transduction that occurred. The results of such an experiment are shown in Table XXX. Note that gene B has been used as an unselected marker. Of those cells that were transduced to A^+, 56 percent also acquired the C^+ marker, and 68 percent the D^+ marker. We can tentatively say that markers A, C, and D are relatively close together, but what is their order? Sixty-six per cent of the C^+ transductants were also carrying the D^+ gene but only 21 percent were A^+. Thus, we can conclude that gene C is closer to D than it is to A, so the order must be A-D-C. Seventeen percent of the E^+ transductants were also carrying the A^+ marker, but only 7 percent were transduced to C^+. Thus, the E gene is closer to A than it is to C, so it must be to the left of the group ADC. Nearly all (96%) of the cells transduced to E^+A^+ also carried the B^- marker, indicating that the B gene was between E and A. The exact order of the genes is, therefore, E-B-A-D-C.

TABLE XXX
GENETIC MAPPING BY TRANSDUCTION*

Selected transductants	A^+	B^-	Percentages of transductants also showing phenotypes: C^+	D^+	E^+
A^+	—	32	56	68	24
E^+	17	25	7	12	—
C^+	21	10	—	66	3
A^+E^+	—	96	27	57	—
A^+C^+	—	28	—	91	8
A^+E^+	88	85	—	89	—

*From Taylor, A. L., and Trotter, C. D.: *Bacteriol Rev, 31*:332, 1967.

Relative distances between markers can be only roughly approximated. For example, the distances between markers D and C and between A and D appear to be shorter than the distance between E and B. Anomalous results occasionally appear in transduction experiments, making such approximations tenuous. For instance, 56 percent of the transductants selected for A$^+$ were also transduced to C$^+$. But in the reciprocal cross in which the C$^+$ transductants were selected for, only 21 percent exhibited the A$^+$ trait. As a rough rule of thumb, transduction frequencies over about 35 percent indicate markers to be less than 0.5 minutes apart,* from 4 to about 26 percent, about one minute apart, and less than 4 percent, 1.5 to 1.8 minutes. Transduction with phage P1 cannot occur between markers of two minutes or more distance because of the limited capacity of the phage head. The molecular weight of the P1 double stranded DNA is 6 x 10^7 daltons, which presumably would represent the maximum amount of bacterial DNA that could be transduced. With the *E. coli* chromosome being 2.8 x 10^9 daltons in molecular weight, the most *E. coli* genome that can be carried by a transducing P1 phage would be about 2.1 percent of the total bacterial genome, or approximately that length transferred by conjugation in two minutes.

Integration

Contrary to the situation in transformation where the donor DNA must be single-stranded before integration, in general transduction the donor DNA appears to be double-stranded on integration. Physically, integration proceeds by breakage and reunion where the size of recipient genome sector displaced is of the order 2 x 10^6 daltons in molecular weight.

GENETICS OF BACTERIOPHAGES

Types of Mutants

Spontaneous mutations were observed to occur in bacteriophages since the time of their discovery. One of the earliest types described are the *host-range* mutations. We know that

---

*Utilizing time of conjugal transfer (Fig. 100) as map distance units.

bacteria can experience mutations that produce resistance towards certain bacteriophages. The basis for the resistance is usually an alteration in the structure of the adsorption sites on the cell surface such that the phage particles can no longer become attached. Bacteriophages can experience mutations, however, that restore to them the ability to infect the bacteria that had previously become resistant. The basis for the enlarged host-range also involves a structural alteration, this time in the phage tail fibers that affords them the ability to attach to previously resistant bacteria.

Bacteriophages generally are counted by the plaque assay in which phage particles and a heavy suspension of a susceptible bacterium are mixed in molten, 0.8% "soft" agar and poured onto an appropriate solid medium. One bacteriophage particle may infect a bacterial cell, which in turn lyses and releases a hundred or more phage particles. These viruses subsequently infect the bacteria surrounding the initially infected cell, and eventually a localized area of cell lysis will form against the confluent growth of the uninfected bacteria (Fig. 108). The areas of lysis are known as *plaques,* and the number of plaques that appear on a plate is related to the number of phage particles in the original suspension. A careful examination of several hundred plaques of a given bacteriophage, such as T4, will reveal slight but reproducible differences in appearance of many of them. Such differences frequently are due to *plaque morphology* mutations that may alter the size, turbidity, or the nature of the center or edge of the plaques.

One of the most intensely studied of the plaque morphology mutants are the *r* mutants of T2. Wild T2 plaques are relatively small, with a definite but turbid center and a fuzzy edge. Mutants possessing the *r* phenotype produce considerably larger plaques with a clearer center and a distinctly sharp edge (Fig. 109). The basis for the *r* phenotype has been explained as follows. Wild T2 bacteriophages impose a characteristic on their host known as *lysis inhibition,* meaning that under certain circumstances bacteria infected by T2 particles do not lyse readily. The circumstance that induces lysis inhibition appears to be superinfection by large numbers of T2 phage particles,

Figure 108. Bacteriophage plaques. A lawn of confluent bacterial growth on an agar plate is punctuated by small areas of lysis where single viral particles infected individual bacterial cells. Within a half hour, these cells lysed and released scores of phage particles, which in turn infected and lysed adjacent bacteria. The lysis spread until it formed the macroscopic plaques seen here.

such as would happen in a developing plaque. The *r* mutants, on the other hand, have lost the lysis inhibition trait and therefore bring about rapid lysis (hence the symbol *r*) to produce larger, clearer plaques with sharp edges.

Other plaque morphology mutants may exhibit greater turbidity *(tu)* or considerably smaller size (*m* for minute). An appropriate word of caution must be included here. Plaque morphologies also may be altered by changes in the strain of bacteria used as the indicator, and by shifts in cultural conditions.

Additional bacteriophage mutants have been isolated that show greater UV or acriflavin resistance, temperature sensitivity, or an altered head structure.

Recombination

In 1946 Delbrück and Bailey, and Hershey working independently, reported that if they simultaneously infected a bacterial culture with two or more different bacteriophage mutants, they observed a few emerging bacteriophage particles that exhibited recombinant phenotypes. That bacteriophages could undergo genetic recombination attracted considerable attention, for their apparent simplicity and ease of handling

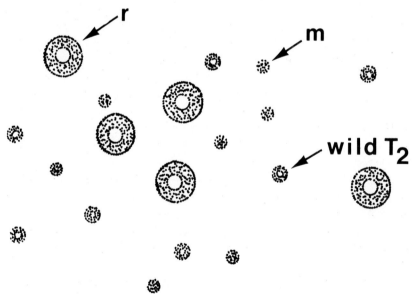

Figure 109. Examples of wild and mutant plaque morphologies in bacteriophage T2.

could not be surpassed by any other system. It was soon possible to construct linkage maps of bacteriophage chromosomes through the use of the observed recombination frequencies between specific markers. However, results of phage crosses began to emerge that defied explanation by classic recombination mechanisms. If one infected a bacterial culture with three phage mutants, say abc^+, ab^+c, and a^+bc, there appeared phage particles that exhibited the phenotype $a^+b^+c^+$. These particles, as can be seen, can be described as having three parents, inheriting the a^+ trait from one, the b^+ trait from another, and the c^+ trait from the third. Such a *triparental cross* was certainly unique by common standards.

In addition, anomalous results began to turn up in the recombination frequencies between markers of known distances. Generally, in higher plants and animals if recombination occurs between markers a and b at the frequency x, and recombination between b and c occurs at the frequency y, recombination between a and c usually is observed to be slightly less than one would expect based on the additive distances a+b and b+c. This phenomenon is known as *interference*. In bacteriophages, on the other hand, recombination frequencies slightly *greater* than one would expect ensued between distant genes. This phenomenon was called *negative interference*.

As a final example of the observations made of recombination in bacteriophages, mention is made of the results of Hershey and Rotman, who analyzed the phage particles that emerged from individual bacterial cells following mixed infections by viruses differing by two markers. In such a *single-burst* experiment, phage particles resulted that possessed the characteristics of one or the other of the parents, as well as the possible recombinant phenotypes. The emergence of parental types AND recombinant types from a single cell would be unexpected, if one assumes that the recombinational event occurs once and at a time before phage DNA replication has proceded to any degree.

Visconti and Delbrück (1953) offered the following hypothesis to explain these and other results not cited here. Recombination in bacteriophages does not occur at a particular time during the lytic cycle, but is a continual series of events that spans the time from early in the infection up to the instant of cell lysis. During the infection period, the phage DNA that has been manufactured enters a pool from which molecules are subsequently withdrawn for encapsulation into phage heads. In mixed infections the DNA's of different superinfecting phages have numerous opportunities of mingle and undergo several *rounds of mating*. It is now easy to understand how a triparental cross may arise. The chromosome carrying the genotype abc^+ undergoes recombination with one of genotype a^+bc to result in the recombinant type a^+bc^+, which in turn recombines with a third chromosome of genotype ab^+c. The recombinant genotype $a^+b^+c^+$ could be the outcome of such a series of cross-over events.

Negative interference was also explained by the Visconti-Delbrück theory, for with two markers that are relatively distant, repeated rounds of mating will lead to greater opportunities for recombination between the markers than would normally be allowed according to their distance.

Probably the most graphic demonstration of the occurrence of repeated recombination in bacteriophages was found by Levinthal and Visconti, who mixedly infected host bacteria and then imposed conditions that inhibited their lysis. The bacteria were thus prevented from lysis for up to 80 minutes. At regular intervals, the workers lysed the bacteria and assayed the resulting phage particles for recombinant types (Fig. 110). A rise in the proportion of recombinant types begins early in the infection cycle, but the curve gradually levels off and approaches a plateau which it has been suggested simulates a level of genetic equilibrium. Genetic equilibrium is a phenomenon seen in many genetic systems and is the level where a balance is struck between proportions of various alleles. That is to say, every population whether bacteria or rabbits carries certain numbers of specific mutants. The relative proportion of the mutant types compared to the wild types depends on the rate at which the mutations occurs, on selective pressures acting against the mutants and on the back mutation rate. After many generations, a level is soon found where the forces that tend to eliminate the mutants are balanced by those that bring about their appearance. What is remarkable is that this phenomenon can be at least partially demonstrated in a single test tube of phage-infected bacteria in a matter of hours, whereas it would take years to do it with a colony of rabbits or other animals.

Fine Structure Mapping

Fourteenth and fifteenth century cartographers continually strove to fill the empty gaps that appeared on their world maps. Similarly, geneticists have tried to account for as much of an organism's genome as is possible through various techniques. Some geneticists have not been as content with merely filling in the larger gaps as they have in determining the topography of very small sectors, such as would occupy the region of a single gene. One such worker was Seymour Benzer, who was able to

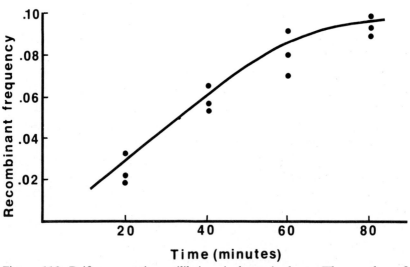

Figure 110. Drift to genetic equilibrium in bacteriophage. The number of recombinant bacteriophage in a mixed infection rises steadily in lysis-inhibited host cells, but a plateau appears to form between 60 and 80 minutes after the infection was initiated. (From Leventhal, D. and Visconti, N.: *Genetics, 38:*500, 1953)

resolve a region of the T4 bacteriophage chromosome, representing about one percent of the total genome, into over 300 divisions.

The sector that attracted Benzer's attention was that which controlled the *r* plaque morphology phenotype discussed earlier. Three distinct regions of the T4 genome are involved with the *r* phenotype. They are designated r_I, r_{II}, and r_{III}, and are distinguishable by their phenotypes on two indicator bacteria. The *r* mutants originating in the r_I region form *r*-type plaques on both *E. coli* strain B and strain K12(λ), whereas the r_{II} mutants form *r* plaques on strain B but no plaques at all on K12(λ). T4r_{III} mutants form *r* plaques on strain B, but wild type plaques on K12(λ). The fact that the r_{II} mutants failed to form any plaques on K12(λ) offered Benzer the sensitive assay system necessary to detect the very rare occurrence of wild recombinant types that would be formed when bacteria were multiply-infected with two r_{II} mutants.

Benzer isolated about 2400 r_{II} mutants of bacteriophage T4. These he separated into two groups according to whether or not they showed significant trends towards reversion to the wild type. Those mutants that failed to show reversions were assumed to be due to deletions. The extent of each deletion was determined by carrying out a small number of pair-wise crosses between the deletion mutants, the result of which was the construction of a genetic map of the r_{II} region. The map was divided into forty-seven segments, the boundaries of which were defined by the limits of the deletions. A simplified version of such a map is shown in Figure 111.

The second, larger group of r_{II} mutants that showed reversions was assumed by Benzer to be composed of point mutants. These were crossed with a few key deletion mutants in order to determine the approximate location of the point mutations. That is to say, if plaques failed to form on K12(λ) when mixedly infected with a known deletion mutant and an unknown point mutant, the point mutation probably was located within the limits of the deletion (Fig. 112). Eventually Benzer carried out crosses with all of the over two thousand r_{II} point mutants with the few (about 40) deletion mutants and

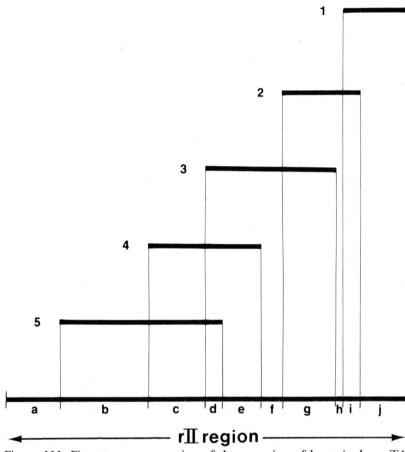

Figure 111. Fine structure mapping of the r_{II} region of bacteriophage T4. Examples of the locations of five deletion mutations in the r_{II} region are shown, together with the sectors that define their boundaries (a through j). (Modeled after Benzer, S.: *Proc Natl Acad Sci USA, 47:*403, 1961)

thereby was able to assign a particular segment for each point mutation. Finally, pair-wise crosses were conducted between the members of each of the segments in order to map their specific location within each segment.

It is important to recognize that if Benzer had not used the deletion mutants to determine the general segments in which the point mutations had occurred, he would have had to carry

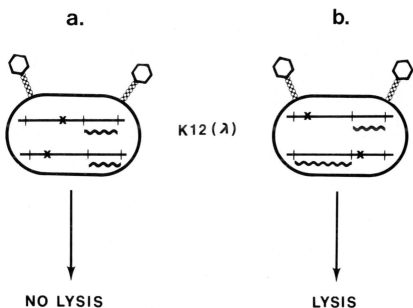

a.   **b.**

K12 (λ)

NO LYSIS   **LYSIS**

Figure 112. Complementation between the two cistrons of the r_{II} region. On one instance (a.) two r_{II} mutants infect a nonpermissive host cell, but both have suffered the mutation in the same cistron, resulting in the termination of the infection. In instance (b.), the mutations have occurred in different cistrons, in which case each mutant genome can still produce the product of the functional cistron. The genomes thus complement one another and the infection is completed.

out nearly five million pair-wise crosses to obtain the same information. As a result of this work, Benzer mapped the locations of some 308 point mutations within the r_{II} region. Assuming the mutations to be equidistant, this represents a resolution of some three to six nucleotides.

Two additional factors make up important parts of Benzer's work. Obviously there was considerable duplication among Benzer's mutants, in that over 2000 mutants appeared to reside in but 308 sites. This quantity of data enabled Benzer to determine whether the observed point mutations occurred randomly throughout the r_{II} region. The fact of the matter was, they did not. There were some sites that showed enormous frequencies of mutation, whereas others experienced but a few mutations. More significant was the comparison between the rates of appearance of spontaneous point mutations and those induced by various mutagens. Certain sites were more susceptible to the action of nitrous acid, for example, and others mutated more frequently under the influence of 5-bromouracil. Benzer referred to those sites that exhibited high mutational activity as *hot spots,* and suggested that the nonrandom mutability observed in his experiments reflected the nucleotide composition of the sites.

As discussed in Chapter Four, specific mutagens invoke predictable changes in nucleotide sequences; 5BU induces AT to GC transitions, for example. Thus, sites rich in AT pairs would be particularly vulnerable to 5BU mutagenesis. But in addition, the genetic code allows an organism to absorb a certain number of nucleotide substitutions (See Fig. 27), depending on the original nucleotide triplet. These two nonrandomly operating factors, susceptibility of certain genetic regions towards specific mutagens, and the capacity of an organism to absorb limited nucleotide alterations, are what apparently account for the uneven distribution of mutations in the *r*$_{II}$ region of bacteriophage T$_4$, and presumably in all genomes.

The second feature that was brought to light by Benzer's work was the discovery that the *r*$_{II}$ region actually consists of two independently operating sectors. That is, the region is made up of two *cistrons,* as Benzer named them, each of which is responsible for the formation of a specific functional polypeptide, and both of which are required for plaque formation on K12(λ). Thus the *r*$_{II}$ phenotype is due to the inability of the mutant phage to produce one or both of the polypeptides.

Evidence for the existence of the cistrons was born out of experiments with the *r*$_{II}$ mutants in which mixed infections were carried out in the nonpermissive host, K12(λ). If an *r*$_{II}$ mutant bacteriophage comes in contact with a K12(λ) cell, normal attachment and DNA penetration occurs, but the infection is prevented from completion. If two *r*$_{II}$ mutants simultaneously infect the K12 host, lysis of the host will depend on whether each of the phage genomes can supply that polypeptide which the other lacks. That is to say, suppose one of the mutants (Fig. 112) has suffered a mutation in the A cistron and is unable to produce its product, but it can still produce the other product. The second mutant may have experienced a mutation in the B cistron, but is still capable of producing the A product. The two genomes thus complement one another and will bring about the lysis of the host cell.

Notice that we are not referring to recombination, for the phage particles that are released from such a *complementation test* still bear mutant genotypes. Recombination is possible, but its occurrence is considerably less frequent than complementation. If the mutants complement one another, the lytic cycle will be completed whether recombination that leads to a wild genotype occurs or not.

The complementation test has been very useful in quickly determining whether conditional lethal mutations have occurred in the same functional unit or not. Edgar and Epstein managed to isolate a number of temperature-sensitive mutants of bacteriophage T2. By pair-wise mixed infections at the nonpermissive temperature, they were able to identify over sixty different functional regions on the phage chromosome. Nonsense *(amber)* mutations can also be tested by the complementation procedure.

Operationally the complementation test can be carried out as follows (Fig. 113). Soft agar is prepared as for ordinary phage assays. To the soft agar, maintained molten at 47C, is added one drop of an overnight, aerated culture of the nonpermissive host, followed by approximately 10^8 particles of one of the phages to be tested. The agar is thoroughly mixed and poured over the bottom agar and allowed to harden. With the aid of a grid pattern on the plate, one drop containing 10^7 particles of the other mutants to be tested is placed on the agar surface. Pipets, stirring rods or applicator sticks may be used for spotting. The plates are incubated and later examined for extensive lysis of the bacterial lawn at the spotting sites. Extensive lysis indicates that complementation had occurred.

If other conditional lethal mutants are to be used, such as temperature-sensitive ones, a permissive host is used as

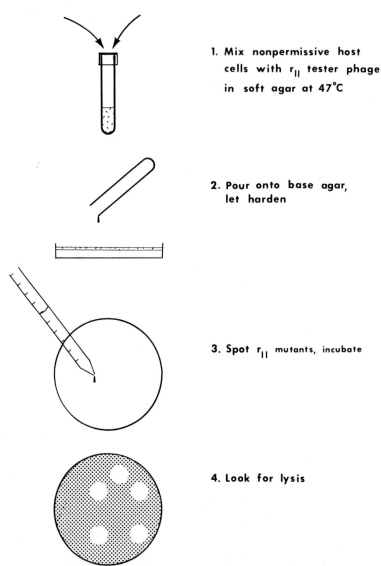

1. **Mix nonpermissive host cells with r_{II} tester phage in soft agar at 47°C**

2. **Pour onto base agar, let harden**

3. **Spot r_{II} mutants, incubate**

4. **Look for lysis**

Figure 113. Steps in the complementation test. Molten soft agar is seeded with tester r_{II} phage mutants and K-12(λ) indicator bacteria and poured over nutrient agar base layer. Unknown r_{II} mutants are then spotted onto the plates, and the plates are incubated. Evidence of complementation is the presence of extensive lysis. An occasional single plaque may appear as the result of recombination, but these are discounted.

indicator, but the plates are incubated under nonpermissive conditions, e.g. elevated temperatures.

Transcapsidation and Phenotypic Mixing

If two different but related bacteriophage particles, such as T_2 and T_4, simultaneously infect a bacterial cell, during the course of the maturation process it is possible that the genome of one of the phages may become encapsulated into the head of the other. This phenomenon, once known as phenotypic mixing (see below) is now referred to as *transcapsidation*. Transcapsidation can lead to some very curious observations, for as one can understand, a phage particle as described above may carry the genetic information of one of the particles, but it expresses the antigenic and host range characteristics of the other.

For example, suppose one infects a bacterial suspension with equal numbers of T_2 and T_4 bacteriophages. A certain number of the resulting progeny will have experienced transcapsidation. Let us say that 10 percent of the progeny consist of T_2 genomes in T_4 heads ("T_2/T_4"). If the progeny from this infection were to be assayed on a mixed indicator consisting of *E. coli* strains B and B/4, *clear* plaques would signal T_2 phage, and cloudy plaques, T_4. If the same lysate were assayed on B/4 alone, any plaques formed would be due only to T_2 phage. Thus, the number of clear plaques on both indicators would be expected to be equal under normal circumstances, but it has been observed that the number on the mixed indicator exceed the number on the B/4 (Table XXXII). The T_2/T_4 particles are capable of attaching to strain B, and the resulting progeny can infect and lyse both indicator strains. These particles cannot, however, attach to and infect strain B/4.

In mixed infections involving two different viruses, occasionally it is observed that some of the issuing viral particles exhibit phenotypic characteristics of both parents. They may carry antigenic determinants of both parents, or they may exhibit the host range of both. On further study it will be found that the

TABLE XXXI
EXAMPLE OF PLASMID CLASSIFICA-
TION

Group	Examples of members
F	F
	ColV2, ColV3-K·30
	F'-lac, ColB1, ColB2
	R1
I	ColIa-CA53
	R64
	R144

genotype of these particles is purely that of one or the other of the parents, but during the assembly of the virions, the protein components of both viruses become mixed so that a phage particle may possess, for example, tail fibers of both infecting particles. As you recall, the tail fibers are involved with host specificity. Thus, such a viral particle will exhibit the host range of both viruses, but its progeny will express the host range specifically denoted by the genotype it possesses. This phenomenon is called *phenotypic mixing.*

PLASMIDS

Extrachromosomal genetic elements of bacteria, capable of independent replication and usually dispensable to the bacteria are known as *plasmids.* Most plasmids such as the F-factor can induce their own transfer through conjugation, in which case they are referred to as *transmissible plasmids, conjugative plasmids,* or *conjugons.* Plasmids also can be transferred from cell to cell via transduction or transformation. Plasmids frequently impart some new phenotypic trait upon the cells carrying them, although through improved electron microscope and other techniques more and more plasmids are being discovered that appear to be *cryptic.* A cryptic plasmid is one that appears not to have an effect on its host's phenotype. Examples of host traits induced by plasmids are bacteriocin and toxin formation, lysogeny, fertility and most significantly, multiple drug resistance. Some plasmids are known to become integrated into the chromosome of the host, in which case they have been given the special name of *episome* by Wollman and Jacob. The λ prophage and the integrated (Hfr) state of the F-factor are examples of episomes.

Classification

The virus-like nature of plasmids suggested that they might be classified according to host, but it has recently been shown that the transmission of many plasmids can cross generic boundaries.

Plasmids usually can be arranged into *incompatibility groups* according to their inability to coexist stably in the same host cell. For instance, the fertility plasmid F cannot stably occupy the same host with plasmid ColV2 at the same time. Thus, F and ColV2 are somehow related and belong to the same incompatibility group. The basis for incompatibility is complex and not fully known, but some suggestions as to its nature have been brought forth. It appears that plasmids may occupy specific sites on the cell membrane. These sites are necessary for the equitable segregation of daughter plasmids among the host's progeny in a manner similar to the distribution of the host chromosomes (Chapter Two). Plasmids that show incompatibility would occupy the same sites, so that if a ColV2 plasmid enters a cell already carrying F, the former is prevented from establishing a stable residence. The superinfecting plasmid is not destroyed, however, but remains in the cytoplasm of the host, waiting to occupy a free maintenance site. The cytoplasmic element usually cannot replicate, and as a result is inherited linearly by progeny cells. Such plasmids have been known to express themselves phenotypically, however, as in the example of an F'-lac particle in an incompatible Hfr strain. The F'-lac plasmid directs the formation of β-galactosidase for several generations, even though it has not been allowed to replicate or become integrated into the host chromosome. Oddly enough, incompatible plasmids occasionally can undergo recombination with one another that in effect rescues at least part of the superinfecting genome and allows for its establishment in the host cell line. Compatible plasmids, i.e. unrelated and stably occupying the same cell may also undergo recombination.

Another possible mechanism for incompatibility may involve the formation of a specific repressor by the resident plasmid. The presence of the repressor prevents the establishment of a related superinfecting plasmid.

Table XXXI shows several plasmids arranged into incompatibility groups.

TABLE XXXII
TRANSCAPSIDATION IN BACTERIOPHAGE*

Type	Number Plated	Plaque numbers and morphologies Indicators	
		B + B/4	B/4
Totals (observed)	100	55 cloudy 45 cloudy	45 clear
T2	45	45 clear	45 clear
T4	45	45 cloudy	0
T2/T4	10	10 clear	0

*A mixture of bacteriophages that presumably consists of equal numbers of T2 and T4 bacteriophage particles is assayed on the mixed indicator B + B/4 and on B/4. T2 should form clear plaques on both indicators, while T4 should form cloudy plaques on B + B/4 and of course no plaques on B/4. However, for every 100 bacteriophage particles plated, 55 show clear plaques on B + B/4 and only 45 clear plaques form on B/4. Forty-five cloudy plaques form on B + B/4.

The explanation for the anomolous results is that 10 percent of the phage particles experienced transcapsidation on maturation, and consist in our example of T2 genomes in T4 capsids. These particles can attach to and infect only the B cells, thus exhibiting the T4 host range, but on lysis of the bacteria, T2 phages are released. The latter then form the clear plaques on the mixed indicator, B + B/4.

Another barrier to the cohabitation of two related plasmids in a single bacterial cell is *entry exclusion*. This phenomenon appears to be independent of incompatibility described above, and is somehow associated with the bacterial cell surface. Patterns of entry exclusion change according to the physiological state of the cell, or after chemical treatment known to alter the cell surface. Entry exclusion is plasmid specific in that, for example, if a donor bacterium carries both an R and a Col plasmid and a recipient carries R, only the R DNA will be prevented from entering the recipient cell on conjugation of the pair.

Types of Plasmids

F was the first plasmid to be discovered, and since that time hundreds of various extrachromosomal elements have been identified in bacteria. Of particular interest are the plasmids responsible for drug resistance in many genera of bacteria. Since their discovery is of some historical as well as practical importance, we will devote some time to them.

Resistance Plasmids

The high incidence of shigellosis in Japan following World War II was countered by the intense use of sulfanilamide (SA) and its derivatives, but by 1950 the occurrence of sulfanilamide resistant *Shigella* rose nearly eight-fold. Other antibiotics, such as streptomycin (SM), chloramphenicol (CM), and tetracycline (TC), soon were brought into use. In 1952, a stool specimen from a dysenteric patient in Kyoto yielded a strain of *Shigella* that was resistant to three antibiotics, SM, TC, and SA. Within a few years the isolation of such strains was commonplace, and by 1955 strains began to appear that were resistant to all four drugs, SM, TC, SA, and CM. The acquisition of multiple drug resistance could be attributed to the occurrence of independent mutations, or to cross-resistance. While these possibilities were being studied, the observation was made in 1959 that many patients with multiple-resistant *Shigella* in their stools also carried strains of *E. coli* with similar or identical resistance patterns. Genetic transfer was suspected, and it soon became clear through work of Akiba and independently by Ochiai in 1959 that the drug resistance could be transmitted among various enteric bacteria by conjugation.

Conjugal transfer of drug resistance appeared to be independent of the presence of the F plasmid, and in 1960 Watanabe and Fukasawa came to the conclusion that the observed multiple drug resistance was due to the presence of a new class of transmissible plasmid, the resistance or R plasmids. The R plasmids are not confined to the enteric bacteria, for they are found in many genera responsible for clinical infections: *Staphylococcus, Pseudomonas, Yersinia, Klebsiella,* and *Neisseria,* as well as in saprophytic yeasts and actinomycetes.

The list of agents against which plasmids afford resistance to their host is also broad, including in addition to the ones cited above, penicillin and its derivatives, and viomycin, gentamicin, kanamycin, spectinomycin, and the heavy metals mercury and cadmium.

TABLE XXXIII
ENZYMES PRODUCED BY R-PLASMIDS*

Enzyme	Substrate
Acetylase	Chloramphenicol, kanamycin
Adenylase	Streptomycin, spectinomycin
Phosphorylase	Streptomycin
Penicillinase	
(β-lactamase)	Penicillin

*From Novick, R. P.: *Bacteriol Rev, 33:*210, 1969.

Most instances of R plasmid-mediated drug resistance can be traced to a specific plasmid gene that is responsible for the formation of a detoxifying enzyme. Table XXXIII lists several enzymes produced by R plasmid genomes and the antibiotic-substrates attacked by them. In other cases, notably in tetracycline resistance, the target of the plasmid enzyme is not the antibiotic molecule, but the mechanism responsible for the drug's uptake by the bacterial cell. Details of this phenomenon are not at all clear, however.

Mechanisms of resistance to inorganic ions attributable to plasmids are not completely known. Two, possibly three, genes are involved in cadmium resistance in *Staphylococcus,* and are found on the same plasmid that also affords the bacteria resistance towards penicillin, lead, and bismuth. The mercury resistance plasmid of *E. coli* appears to have the ability to reduce mercuric ion (Hg^{+2}) to the volatile and less toxic metallic form, Hg^0. Lastly, certain R plasmids in *E. coli* and *S. typhimurium* can confer increased resistance towards ultraviolet light to these organisms. Interestingly, other R plasmids have been shown to increase the cells' sensitivity towards UV. It has been suggested that plasmid intervention of DNA repair is somehow responsible for these responses.

Origins of R^+ Bacteria

R^+ bacteria have been isolated from geographically remote human and animal populations that presumably have never been exposed to selective levels of antibiotics. This strongly suggests that R plasmids existed before the widespread use of antibiotics in the last thirty years, and they are probably common in nature, though in small numbers. Large numbers of R plasmid-carrying strains appear rapidly following the introduction of antibiotics, as in the case of their original discovery in Japan. Several bacteriologists have reported that there is almost a direct correlation between the amount of antibiotics used in a given hospital and the proportion of R^+ bacteria in their sewage. Significant increases in the numbers of R^+ bacteria have also been detected in domestic animals and cultured fish that have been treated with high levels of antibiotics as routine dietary supplements. While many of the bacteria isolated in these cases are pathogenic only for their animal hosts, others, particularly strains of *Salmonella,* are known human pathogens. Furthermore, one must keep in mind the promiscuous interspecific and intergeneric transfer of R plasmids that has been observed.

Tons of streptomycin, tetracycline, and other drugs have been used on agricultural crops to retard field infections, and on meat and poultry to control spoilage. The increased incidence of R^+ bacteria appears to be the consequence of the broad-brushed medical and nonmedical dissemination of antibiotics. It has come to the point that R^+ strains of bacteria are developing faster than man can find new antibiotics to combat them.

Col Plasmids

Colicins are a class of highly specific antimicrobial molecules called *bacteriocins.* Colicins are specifically produced by and are active against strains of *E. coli* and other members of the enteric bacteria. Their formation is under the direction of a type of plasmid known as a *Colicinogenic factor* or *Col.* A number of different Col plasmids have been discovered, most of which are transmissible and fit into the F-like incompatibility group (Table XXXII) and a few in the I-like group. There is also a group of nontransmissible Col plasmids, exemplified by ColE1, ColE2, and ColE3. Many of the colicins, such as K and D, appear to be fragments of bacteriophage particles, and it is therefore proposed that at least some Col plasmids may be defective phage genomes that are capable of directing the formation of components of the bacteriophage particles that are still lethal to the bacteria. Bacteriocins from other genera, such as pyocins from *Pseudomonas* and lethal particles from *Bacillus* species, also appear to be phage fragments, but their origins have not been connected with any sort of plasmid-like entities. Susceptibility to bacteriocins rests on the presence of specific receptor sites on the bacterial cell surface, and mutations that alter these sites may lead to bacteriocin resistance.

The mode of action of bacteriocins varies widely. In the case of colicins E1, D and K, ATP formation is blocked, disrupting all pathways dependent upon this molecule. Colicin E2 and megacine C from *B. megatherium* block DNA synthesis, and colicin E3 stops protein synthesis.

Metabolic Plasmids

Many soil bacteria have the capacity to decompose aliphatic and aromatic hydrocarbons such as hexane, benzene, catechol, phenol, naphthalene and camphor. Camphor, for example, (Fig. 114) is oxidized by *Pseudomonas putida* to acetate and isobutyrate. The ability to degrade these compounds is frequently controlled by plasmids that are transmissible either through conjugation or through transduction. A remarkable aspect of these extrachromosomal particles is that they appear to possess the entire complement of structural genes necessary for the formation of all the enzymes involved in the pathway, in addition to those that control autonomous replication and transfer. Such clustering of common genes in the chromosome of *Pseudomonas* is infrequent, leading to the supposition that if the plasmids are of bacterial chromosome origin, they must have come from another genus.

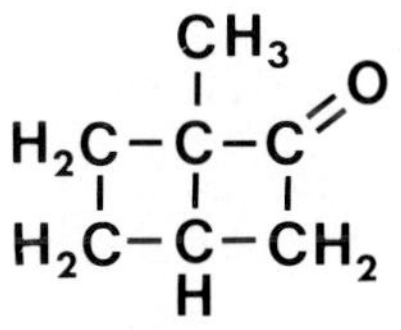

$$H_2C-\overset{\overset{\displaystyle CH_3}{|}}{\underset{\underset{\displaystyle H}{|}}{C}}-\overset{|}{\underset{|}{C}}\overset{O}{\diagup}$$
$$H_2C-C-CH_2$$

Camphor

Figure 114. Structure of camphor, one of many hydrocarbons attacked by enzymes produced by plasmid genes.

Replication

All plasmids so far tested consist of closed, circular double-stranded DNA. Evidence for both the Cairns closed circle mode of replication (Fig. 15) and the rolling circle model (Fig. 18) has been elicited for the plasmids. Although nonintegrated plasmids replicate autonomously, it appears that at least some bacterial genes are involved in plasmid replication. Most plasmids follow a regulated replication cycle that is synchronized with host chromosome replication and which results in a constant, small number of plasmid copies per cell (*stringent* replication). In other instances, control of plasmid replication may be *relaxed,* and plasmid synthesis continues randomly and unabated, regardless of the rate at which the host chromosome replicates. Relaxed replication may result in the presence of twenty or more plasmid copies in a bacterial cell.

Transfer and Genetics of Plasmids

Mention was made earlier of the distinction between transmissible and nontransmissible plasmids. The former are genetically capable of bringing about their own transfer to recipient bacteria via conjugation, whereas the latter type must depend upon other means of transportation. Plasmids not capable of mobilizing their own transfer can be taken up by transducing bacteriophages, or can become associated with transmissible plasmids, such as F or R types, and accompany the latter into recipient cells. Transmissible plasmids induce the formation of pili, which appear to recognize and attach to recipient bacteria, and possibly to act as conduits of plasmid DNA.

Some of the mechanics of plasmid transfer were covered in Chapter Six in connection with conjugal transfer in Hfr strains. While fine details are not clear, it appears that the plasmid DNA is unwound and a single strand is transferred into the recipient cell. Here replication occurs to restore the circular, double-stranded configuration of plasmid DNA. At least eleven genes on the F plasmid are implicated in its transfer, but the nature and the mechanisms of their products are just becoming understood.

The genetic make-up of the plasmids is not well known except for some of the F and R elements. In the instance of the resistance plasmids, they appear to consist of two regions, the *Resistance Transfer Factor,* or *RTF,* and the *resistance determinants.* The RTF portion carries the genetic information necessary for the replication, mobilization, and transfer of the R plasmid, whereas the resistance determinant portion consists of the complement of genes that afford the host cell the multiple drug resistance characteristic of the R plasmid. Being DNA, R plasmids are subject to mutations, examples of which are: loss of resistance for a specific antibiotic, or an increase in resistance level, loss of transmissibility, either through lack of pilus formation or through other mechanisms, and loss of replication ability. Deletions are known to occur in plasmids, as are nonsense and temperature sensitive mutations. These can be induced by the common mutagens discussed in Chapter Four.

Recombination between plasmids has been observed in a few cases and is assumed to occur by mechanisms already discussed. Composite plasmids containing genes from a variety of sources, F, R, and Col plasmids, for example, have been assembled through recombination. As one might expect, the demonstration of recombination between plasmids runs into problems of incompatibility which tends to select against one of the partners. As plasmids become more distantly related, incompatibility becomes less of a factor, but recombination also becomes less frequent because of fewer regions of homology. Recombination between the host chromosome and the plasmid is also possible.

Certain plasmids may mobilize the transfer of host chromosomal material. Under certain conditions, the presence of the R plasmid will induce the transfer of bacterial genes to a recipient cell. It is not clear whether the R factor is physically associated with the host chromosome at the time of bacterial chromosome transfer, as is the case with the Hfr state. There has been the suggestion that the nonintegrated F plasmid may still bring about the transfer of bacterial genes, which would partly explain the recombinants that result from F^+ x F^- crosses.

Plasmid Interactions

The previous discussion of compatibility alludes to the fact that plasmids frequently interact with one another when carried by the same host bacterium. Some R plasmids inhibit the expression of fertility when carried by F bacteria (known as fi^+ *plasmids*, for fertility inhibiting) whereas other R plasmids (fi^-) do not appear to affect the behavior of the F plasmid. *Phage restriction* is also observed with certain R plasmids in which the DNA of certain superinfecting bacteriophages is destroyed on entering a cell carrying a restricting R plasmid. Restriction appears to involve the specific formation of a unique endonuclease.

In light of the foregoing information, one may wonder how reliable phage typing patterns are, for it appears that susceptibility to bacteriophage activity may be altered by the mere acquisition of a restricting plasmid.

Curing

Various bacterial plasmids may experience rare, abnormal replication that leads to one of the daughter cells missing the plasmid and of course, the associated phenotype. This phenomenon can be induced in the laboratory by a number of treatments and is known as *curing* or *elimination*. Compounds such as the acridines and ethidium bromide can often cure a bacterial cell of a resident plasmid, as can thymine starvation in *thy*⁻ auxotrophs. The ability to cure a bacterium of a plasmid is frequently corroborating evidence for the presence of the extrachromosomal element, but failure to cure is not indicative of nonplasmid genes, for many plasmids are not curable. The mechanism of curing seems related to damage incurred in the plasmid DNA that prevents its normal replication.

Gene Amplification

It recently has been made possible to incorporate heterologous DNA into plasmid molecules *in vitro*. This can be done by first opening up the circular plasmid molecule with restriction endonucleases (discussed earlier in relation to host controlled modification). The opened plasmids have "sticky ends" that will polymerize with other DNA molecules with similar ends, regardless of source. The stickiness is due to a property of the restriction enzymes that leaves complementary ends. The opened plasmid DNA may have the following terminal base sequence:

$$AATT\ldots\ldots\ldots$$
$$\ldots\ldots\ldots TTAA$$

This molecule will now polymerize with any other linear

fragment of DNA that ends with the same sequence of bases resulting from restriction endonuclease treatment:

AATT.........AATT################
.........TTAA ############### TTAA

The formation of such fragments depends on the occurrence of the sequence AATT in their nucleotide composition, and such sequences apparently are common in DNA's of procaryotic and eucaryotic origin. It has thus been possible to construct circular, biologically active plasmids that have integrated within their structure genes from bacteria and other sources. A host bacterium thus made to carry such composite plasmids may, under proper conditions, express those genes added to the plasmid.

A possible application of the creation of hybrid plasmids is the ability to isolate large quantities of the products of the added genes. This is made possible through the use of strains of bacteria that exhibit relaxed plasmid replication. These strains may contain as many as twenty copies of the plasmid, and may demonstrate a dose effect in which quantities of the gene product would be produced in proportion to the number of plasmids present. Thus, a bacterium carrying twenty copies of the β-galactosidase gene will produce twenty times the normal quantity of that enzyme. This technique is known as *gene amplification.*

Discussions are now going on regarding applications of composite plasmids in human medicine. If a satisfactory analog to bacterial plasmids could be found for human cells, it might be possible to "treat" certain metabolic diseases by inserting operational genes into malfunctioning cells. This endeavor is known as *genetic engineering.* The genomes of certain oncogenic (tumor-producing) viruses have some attributes of such a plasmid analog, and the potential hazards of handling such molecules invariably come up in every discussion of genetic engineering. Risks are even present in certain experiments with the common intestinal bacterium *E. coli.* At least one international conference has already debated the risks of such experiments, and proposed placing a moratorium on certain types of work until physical and biological safeguards, usually reserved for highly virulent pathogens, can be developed. An example of a biological safeguard would be using strains that could not survive outside laboratory media, such as multiauxotrophs or drug-dependent varieties.

Chapter Eight ─────────────────────────────

VIRUSES OF EUCARYOTES: FROM FUNGI TO MAN

T HE EUCARYOTES REPRESENT an immensly diverse range of hosts for viruses, commencing at one end of the scale with the eucaryotic microorganisms, continuing through the invertebrates, the lower vertebrates and completing the spectrum with the mammals, including man. Also in our consideration are members of the plant kingdom, from the simple, single-celled algae to the more complex, higher forms. Members of each of the major levels on our scale play host to one or more viruses. Yet in spite of the diversity of the hosts, the viruses of the eucaryotes are morphologically homogeneous, falling into but two geometrical classes: *icosahedral* (or cubic) and *helical*. The icosahedral viruses exhibit a central nucleic acid core, surrounded by a protein *capsid,* the two combining to form the *nucleocapsid.* The nucleocapsids of helical viruses consist of hollow, helical arrangements of *capsomeres* among which the nucleic acid is threaded. Figure 115 depicts these two fundamental morphologies. In certain viruses of either type, the nucleocapsid may be surrounded by a loose membrane or *envelope.* Herpes and influenza viruses are examples of enveloped viruses.

Table XXXIV lists some basic characteristics of the nucleic acids of several viruses of eucaryotes. Note that all four possible forms of nucleic acids are represented: double- and single-stranded RNA, and double- and single-stranded DNA. The range of genome sizes spans nearly two orders of magnitude, with broad bean mosaic virus possessing a mere three genes, to vaccinia, which has the genetic capacity of over 260 genes.

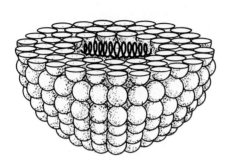

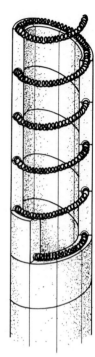

Figure 115. Fundamental morphologies of the viruses of the eucaryotes. Left side: Icosahedral or spherical virion consisting of many capsomeres. The nucleic acid is found in the center of the particle. Right side: Helical morphology in which the nucleic acid is intercalated between capsomeres.

TABLE XXXIV
CHARACTERISTICS OF THE NUCLEIC ACIDS OF SOME VIRUSES OF EUCARYOTES*

Virus	Type of nucleic acid	Molecular weight (Millions of daltons)	Genetic capacity (genes)[†]
Broad bean mosaic	ss RNA	1	4
Tobacco mosaic	ss RNA	2	8
Reovirus	ds RNA	15	30
Wound tumor	ds RNA	16	32
Minute mouse	ss DNA	1.5	6
Polyoma	ds DNA	4	8
Cauliflower mosaic	ds DNA	5	10
Adenovirus	ds DNA	23	45
Fowlpox	ds DNA	160	320
Vaccinia	ds DNA	160	320

*Data from various sources.
[†]See footnote for Table XXV.

THE VIRUSES OF THE EUCARYOTIC MICROORGANISMS

Whereas our knowledge of the viruses of the bacteria is indeed extensive, our understanding of the viruses of the eucaryotic microorganisms is comparatively primative. Information on the viruses of the fungi, algae, and protozoa is limited for the most part to bare morphological descriptions. Modes of transmission and replication and certainly the genetics of these agents are generally unknown. However, these agents have only been discovered within the last decade or two, and we can thus expect a rapidly expanding interest in them in the future.

The first fungal viruses were found in cultured mushrooms in the early 1950's. Since then, viruses have been identified with members of all of the major classes of fungi. The most striking characteristic of the fungal viruses is their biophysical semblance. They all appear to be icosahedral particles of 30 to 40 nm diameter, and to possess double-stranded RNA.

Only a few of the fungal viruses have been studied extensively. These viruses exhibit a wide range of virulence, but examples of definite lysis of host cells are rare. These agents appear to prefer latent responses in which large numbers of particles accumulate in otherwise healthy-looking hyphae. A plaque assay system similar to that utilized with bacteriophages has been developed for certain *Penicillium* viruses, but it has not been free of problems. It seems that the ability to form plaques is under the control of a relatively unstable genetic determinant in the host, making any plaque assay methods somewhat unreliable.

Since host lysis and release of infectious particles is uncommon in the fungi, the principal mode of viral transmission appears to be through the mechanism of *plasmogamy,* in which the mycelia of appropriate mating types come in contact and exchange cytoplasm.

Evidence has been collected to suggest that some fungal viruses can influence certain physiological characteristics of their host. In most cases the influence is in a negative direction; that is, presence of virus appears to reduce the formation of a given metabolite, or diminish the phytopathogenicity of the host fungus. There is no evidence that features akin to transduction or transfection in bacteria occur in the fungi.

Brief note was made in Chapter Seven of the bacteriophage-like viruses of the procaryotic algae. These viruses appear to be widespread in nature, and can be isolated from scummy ponds and propagated in the laboratory with ease. A few eucaryotic algae, as for example *Chlorella* and *Oedogonium,* have been shown to carry polygonal particles in the cytoplasm that resemble viruses. Cellular changes frequently accompany the particles, such as chromatin aggregation and distortions of the mitochondria. The course of the infection and modes of transmission of these putative viruses are essentially unknown.

The protozoa, of which there are over 30,000 species, are considered single-celled animals. Numerous reports appear in the literature describing virus-like particles within the cytoplasm of various protozoa. In at least one case, infectious particles causing lysis of host cells were isolated from a protozoan, *Entamoeba histolytica.* Serial passage of the virus was demonstrated, indicating intracellular replication of the agent.

It is clear from this abbreviated discussion of the viruses of the eucaryotic microorganisms that little is known of them. In fact, the rationale for including them here is in the hope that they will attract attention and perhaps become the object of additional investigations.

THE PLANT VIRUSES

The first virus ever to be isolated was tobacco mosaic virus (TMV) by Iwanowski in 1892. This Russian botanist apparently did not fully understand his discovery and it remained for others to develop an explanation of these "infectious filterable agents." One of the early pioneers in virology was the Dutchman Martinus Willem Beijerinck. Beijerinck coined the term *virus* for the infectious agents that passed through bacteriological filters, and probably did more to stimulate others to carry out investigations on the viruses than any other man of his day.

There are several hundred plant viruses known, many of them being the causative agents of crop-damaging diseases. Only the phytopathogenic fungi bring about greater economic losses to farmers and orchardmen. The plant viruses exhibit two basic morphological types, icosahedral and helical. Most contain single-stranded RNA. A considerable amount of information has been assembled on the biophysical properties of some of the plant viruses, particularly TMV. This is due in the main to the relative ease with which large quantities of these viruses can be isolated and purified. Witness to this is the 1935 work of Wendell Stanley, who managed to purify TMV to the point of its crystallization.

Attempts to demonstrate recombination in the plant viruses generally have been unsuccessful, mainly because of interference between genomes in mixed infections, the small sizes of the genomes, and high spontaneous mutation rates.

TABLE XXXV
EXAMPLES OF VIRUSES WITH DIVIDED GENOMES*

Virus	*Nucleic Acid Type*	*Genome Molecular Weight (Millions of daltons)*	*Number of Components*
Broad bean mosaic	ss RNA[†]	1	3
Alfalfa mosaic	ss RNA	3.3	4
Tobacco rattle	ss RNA	3.1	2
Influenza	ss RNA	3	5
Reovirus	ds RNA	15	10

*Data for genome size represents the approximate, aggregate molecular weights of the known genome components.
[†]ss = single-stranded; ds = double-stranded.

Table XXXV lists some properties of several viruses. It can be seen that these infectious agents possess the remarkable characteristic of having divided genomes. That is to say, the viral particles appear in nature as two or more distinct components. Most of the components must be present in the plant cell in order for the infection to proceed. In some instances, such as in the case of tobacco rattle virus, one of the particles carries genetic information necessary for nucleic acid replication, but complete virion formation is not possible in the absence of a second particle. In alfalfa mosaic virus, three of its five components must be present for virion formation to occur. Apparently each component carries a limited array of viral genes, and a minimum number of them must be present for the infection to follow its normal path.

One plant virus, tobacco mosaic virus, played an important corroborating role in the elucidation of the genetic code. Because such large quantities of relatively pure TMV could be isolated, it was not long before the sequence of the 158 amino acids of the capsid was determined. It was then possible to correlate predicted changes in the base sequences of the viral RNA induced by various mutagens with observed changes in the amino acids of the capsid. As an example, following nitrous acid treatment cysteine residues appeared in the capsid in positions normally occupied by arginine or tyrosine. Both substitutions are clearly explained by a C→U or an A→G transition. The arginine codon CGU is converted to UGU (=cysteine), or the tyrosine codon UAC is converted to UGC (=cysteine).

Viroids

These are extremely small bits of infectious nucleic acid that appear by themselves to be the cause of certain plant diseases. They are not merely the free nucleic acids of known viruses devoid of capsid, for their molecular weight is a mere 5 to 10 percent of the smallest genomes known. Two examples of diseases thought to be caused by viroids are potato spindle tuber disease and chrysanthemum stunt disease. The mode of transmission of these agents is unknown. There is fragmentary evidence that at least one animal disease, scrapie in sheep, may be caused by a viroid.

THE ANIMAL VIRUSES

The considerable interest focused on the animal viruses stems from their role as the cause of many human diseases and diseases of domestic animals. Many of these viruses can be grown in animal cells cultured in the laboratory. The result of such infections is usually the death of the cells. In addition, some viruses have been shown to be the direct cause of tumors in animals and on occasion malignant changes in cultured animal cells. The induction of these latter changes is known as *transformation,* a misleading choice of terms in that it has nothing to do with bacterial transformation. Virus-induced transformation of animal cells is characterized by the destruction not of the cells but of the cells' regulation of normal growth, in addition to other physiological alterations. Transformed cells multiply unrestrained, forming multilayered accumulations of disarranged cells against the tidy background monolayer of normal cells (Fig. 116).

Viruses that produce tumors in animals are known as being *oncogenic,* and while unequivocal proof for the existence of oncogenic viruses as the cause of some human cancers is lacking, the evidence certainly points to that possibility. There is strong evidence that the genomes of some oncogenic viruses of mice and fowl, for example, are carried by the host genome and are thus genetically transmitted from generation to generation. Cells so infected may be induced with various physical and chemical agents and subsequently release mature

Figure 116. Normal cultured animal cells and transformed cells. Normal cells (top) form uniform monolayers on the surfaces of culture vessels, whereas cells transformed by oncogenic viruses (lower drawing) accumulate in multilayered foci.

virus particles. The tenuous parallel between this phenomenon and lysogeny in bacteria may be made, but further work is needed before the comparison is justified.

Convenient culture and assay systems are not available for most animal viruses. Some animal viruses can be titered by pock assay in which suitably diluted suspensions of infectious particles are placed on the chorioallantoic membrane of chick embryos. Following incubation, infectious centers reveal themselves as visible lesions, or pocks, on the membrane surface, and the number of pocks can be related to the titer of the original virus suspension. With the development of *in vitro* animal cell cultures, particularly cell monolayers, plaque assays analogous to the phage plaque assay have been possible with certain animal viruses. Viruses that lack a focal assay method (based on pock or plaque formation) can be titrated by various tube-dilution methods, such as those based on hemagglutination, or by animal assays based on the appearance of disease or death. However, these latter methods are costly and cumbersome and therefore have discouraged studies on a number of viruses.

Nucleic Acids and Replication

The nucleic acids of the animal viruses occur as both single-stranded and double-stranded DNA or RNA (Table XXXIV). Base compositions of the double-stranded DNA viruses appear to cover roughly the same range as is seen in the bacteria, that is from about 32 to 35 percent G+C for the poxviruses, to a high of 74 percent for certain members of the herpesvirus group. Some uncertainty exists regarding the biophysical nature of the nucleic acid as it occurs in mature virions of certain viruses. It appears that just as in some plant viruses, the nucleic acid of some animal viruses is present in the virion as two or more distinct pieces.

The RNA of vesicular stomatitis virus occurs in several components in the mature virion. Reovirus RNA, which is double-stranded, appears to occur in ten distinct fragments of unique size and base composition. The uniqueness of the

reovirus nucleic acid fragments suggests that they are genetically unique as well. That is, each carries separate genetic
information. This also appears to be true for influenza virus
and is supported by the following genetic evidence. Complete
replication of the viral particles is not possible unless a full set of
RNA fragments is present in the host cell. Influenza virus
produces some "incomplete" virions during the normal infectious cycle that lack one of the RNA fragments, and these
particles are noninfectious.

In other instances, animal virus particles actually possess
multiple copies of their genomes, or are *multiploid*. This
phenomenon occurs in the myxoviruses such as parainfluenza
and Newcastle disease virus, and is a result of *budding*. Budding
refers to the manner by which these viruses are released from
the host cell. Virus core nucleoprotein becomes enveloped in
host cell membrane material (Fig. 117), which frequently

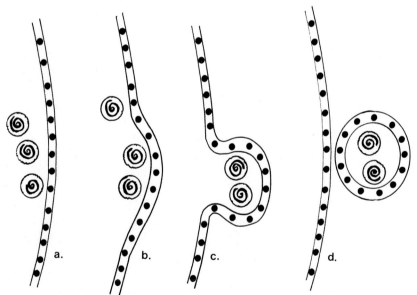

Figure 117. Myxovirus release through budding. Viral nucleocapsids
approach the cell membrane (a.) in which viral-associated antigens have
accumulated as a result of the infection (dark circles). The formation of a bud
is induced (b.) by which two viral genomes (c.) become encapsulated within
the same virion (d.).

results in two or more viral genomes becoming encapsulated in the same infectious particle.

There are many instances where enzymatic activity is associated with purified viral particles. Most of the enzymes appear to be involved with nucleic acids as ribonucleases, deoxyribonucleases, or nucleotide phosphohydrolases. The roles of these enzymes appear to be the replication of the viral genome, the transcription of mRNA, or the modification of host nucleic acids.

Of great interest has been the discovery of DNA polymerase activity in the virions of the leukoviruses, a highly distinctive group of oncogenic agents. What makes this observation immensely significant is that the leucoviruses are RNA viruses, and that the DNA polymerase is RNA dependent. In short, these viruses apparently practice *reverse transcription* in that viral messenger RNA is produced via a DNA intermediate: viral RNA→ DNA→ mRNA. More significantly, the DNA that is produced may become integrated into the host genome as a *provirus,* to be carried through many host cell generations until it is induced. The relationship, if any, between the ability to carry out reverse transcription and the tumor-producing capacity of these viruses is under intensive investigation.

Other examples of enzymes associated with viral particles are the protein kinases, the activity of which is the phosphorylation of serine and threonine residues in certain viral polypeptides. Neuraminidase activity has been known for a number of years to occur in the virions of the myxoviruses. Its activity is to break certain glycosidic bonds in the cell wall of the host.

It is not at all clear why certain animal viruses have specific enzymes as part of their nucleocapsid structure. They apparently are not fortuitous inclusions within the viral particles, for if this were the case, one would expect to find other virus-coded enzymes as well. Until more can be learned of the functions of these enzymes, one can only surmise that they provide greater self-sufficiency to the viruses in the early stages of the establishment of an infection.

Details regarding the replication of animal virus nucleic acids are generally lacking, but there are some observations of specific viruses. Reovirus, which possesses double-stranded RNA, appears initially to carry out asymmetric replication in which the parental double-stranded RNA molecules are conserved while only single-stranded + strands are formed. The released + strands then act as templates for the formation of double-stranded RNA for incorporation into mature viral particles. At the same time, some of the + strands act as mRNA for the formation of viral-specific proteins.

The nucleic acids of certain animal viruses can be extracted with phenol or other reagents and retain infectiousness for animal cells in culture. This is analogous in many respects to the phenomenon of transfection observed in the bacteria (Chap. Seven). As many as 10^8 virus particle equivalents of nucleic acid may be needed to initiate one infected cell. The host range of infectious viral nucleic acid is considerably broader than that of the intact viruses, making it possible to study the replication of certain viruses for which natural host cell cultures are not available. Only a single replication cycle is possible however, for any released mature virions could not reinfect neighboring cells.

Mutations in Animal Viruses

In those viruses that have suitable assay systems to detect them, numerous types of mutations have been observed. These may involve pock or plaque morphology, host range, behavior towards physical or chemical agents, antigenic properties, and other characteristics. Mutations in animal viruses occur spontaneously, or may be induced by any of the mutagens described in Chapter Four. Nitrous acid is commonly used on free viral particles, and 5-bromouracil and 5-fluorouracil have been added to cell cultures infected with DNA or RNA viruses, respectively, to bring about mutations.

The most frequently utilized mutation type in animal virus research is the temperature-sensitive, or *ts,* mutation. These mutations, as you recall, prevent an organism from reproducing at slightly elevated incubation temperatures, say 38 to 40°C, whereas the wild type is capable of doing so.

Every gene of the viral genome that is responsible for a functional protein is theoretically susceptible to a *ts* mutation, thus making it possible to account for most of the viral genome. Among the smaller viruses, such as polyoma, Newcastle, Sinbis, and vesicular stomatitis, over half of the estimated genome has now been identified. For larger viruses, the extent of our genomic knowledge may not exceed 10 percent however.

The isolation of *ts* mutants begins by growing the wild stock at the nonpermissive temperature for the purpose of eliminating mutants already formed and carried in the stock. A clone of wild type viruses is isolated and generally treated with a mutagen, followed by plating in a suitable plaque assay system. Plaques that are formed at the permissive temperature are replated in replica and incubated at both permissive and nonpermissive temperatures. Viruses that form plaques at the lower temperature but not at the higher, are picked for further testing. Detecting stable mutants among the animal viruses has always been a problem, for often high frequencies of reversions or leakiness are encountered. Each isolate must be carefully screened by several replatings at various temperatures to confirm the *ts* character.

It is generally presumed that the basis for the *ts* mutant phenotype is a missense mutation that has resulted in the formation of a heat-labile protein. The association between the mutant gene and its phenotypic expression may not be so direct, however. A *ts* mutant of Semliki Forest virus failed to show RNA polymerase activity when incubated at elevated temperatures, but viral RNA polymerase isolated from infected cells grown at normal temperatures was stable and retained activity in *in vitro* tests at higher temperatures. Apparently the *ts* mutation did not affect the polymerase molecule directly, but possibly its formation.

An example of a more specific mutation in an animal virus is guanidine and HBB resistance in polio. Guanidine hydrochloride, $(NH_2)_2C:NH_2Cl)$, and HBB, (2α-hydroxybenzyl)-benzimidazole, inhibit the multiplication of several RNA viruses by interfering with viral RNA synthesis. Host RNA formation is apparently unaffected. Numerous mutant viruses have been isolated that have acquired resistance toward the effects of these drugs, and in some instances, a dependence on them has been observed.

Another type of mutation of note is one observed in vaccinia and in herpes. Cells infected by wild virions of these agents show an increase in thymidine kinase activity. Certain virus mutants have been discovered that fail to produce the enzyme. The enzymes apparently are specifically coded by the virus genomes for they are immunologically distinct from the host enzyme. Their function is not clear.

Some mutations in animal viruses have exhibited pleomorphism. For example, certain plaque morphology and temperature sensitivity mutants of polio virus possess lowered virulence for primates by intracerebral inoculation. In another instance of greater practical significance, an attenuated mutant of polio is more strongly retained on a cellulose ion exchange chromatography column than is the virulent wild type. If this principle were found to apply to other viruses, it would prove to be a useful general technique in the isolation and purification of appropriate viral strains for the manufacture of vaccines.

Recombination and Complementation

The first observation of recombination in an animal virus was made in influenza by Burnet in 1949. Since then, recombination has been demonstrated in some of the poxviruses, the adenoviruses, and in polio and reovirus. It does not appear to be a general phenomenon, or at least laboratory manipulations have not been adequate to show recombination in other viruses such as Newcastle disease or polyoma.

Recombination is best demonstrated by carrying out mixed infections with conditional lethal mutants, such as temperature sensitive mutants, where wild recombinants form plaques at the nonpermissive condition. Maximum frequencies of recombination vary widely, depending on the type of virus, from less than 1 percent to as much as 50 percent. It has subsequently been shown that the unusually high frequencies of recombination in certain viruses is probably due to divided genomes, and what is

observed is actually the reassortment of nonidentical components rather than true recombination between homologous
nucleic acid molecules. As in the case with bacteriophage
recombinants, suspected recombinant animal viruses must be
subcultured and tested further to confirm their new genotype.

Complementation experiments identical in principle to those
described for bacteriophages (Chap. Seven) can be carried out
with certain animal viruses. Pairs of *ts* mutants are made to
infect simultaneously host cells that are incubated at the
nonpermissive temperature. Cells that are infected with
nonidentical (occurring in different genes) mutant viruses will
lyse and release particles, ultimately resulting in the formation
of plaques. Generally, no plaques will form if the viruses have
experienced mutations in the same gene. Eventually, one can
classify a series of mutant viruses into complementation groups
which presumably define discrete functional genes in the virus
genome. Unfortunately, complementation tests with animal
viruses operationally are not as simple as the spot tests
described earlier for bacteriophages, but they still are widely
used and offer the first step in the genetic mapping of the
animal virus genome.

In those DNA animal viruses studied, nucleic acid replication
appears to occur in pools in much the manner envisioned by
Visconti and Delbrück in bacteriophages. Continuing opportunities for recombination among heterogenic molecules in
mixed infections during the course of the infectious cycle have
also been observed in the animal viruses. In herpes virus (type
1), for instance, in a mixed infection with two *ts* mutants, the
frequency of recombination observed among the emerging
mature virions rose during the infection to reach a plateau after
about eighteen hours.

By contrast, in the RNA animal viruses where recombination
has been observed, the event appears only to take place early in
the infection, perhaps exclusively by the parental RNA. In
crosses with mutants of influenza virus, the proportion of
recombinants that appear eight hours following the initiation of
the infection is not changed significantly over the next twelve
hours. Basically, the same observation has been made with
reovirus.

BIBLIOGRAPHY

CHAPTER 1

INTRODUCTION

Books and Review Articles

Carlson, E. A.: *The Gene: A Critical History.* Philadelphia, Saunders 1966.

Childs, B.: Garrod, Galton and clinical medicine. *Yale J Biol Med, 46:*297, 1973.

Dobell, C.: *Antony van Leeuwenhoek and his Little Animals.* New York, Dover, 1960.

Gardner, E. J.: *Principles of Genetics,* 2nd ed. New York, Wiley, 1968.

Lechevalier, H. A., and Solotorovsky, M.: *Three Centuries of Microbiology.* New York, Dover, 1974.

Levine, L.: *Biology of the Gene.* St. Louis, Mosby, 1973.

Ravin, A. W.: *The Evolution of Genetics.* New York, Acad Pr, 1966.

Research Papers

Beadle, G. W., and Tatum, E. L.: Genetic control of biochemical reactions in *Neurospora. Proc Natl Acad Sci USA, 27:*499, 1941.

Muller, H. J.: Artificial transmutation of the gene. *Science, 66:*84, 1927.

Sutton, W. S.: The chromosomes in heredity. *Biol Bull, 4:*231, 1903. (Reprinted in Peters, J. A. (Ed.): *Classical Papers in Genetics.* Englewood Cliffs, Prentice-Hall, 1959.)

CHAPTER 2

HISTORY OF GENETIC MATERIAL

Books and Review Articles

Bresler, S. E.: *Introduction to Molecular Biology.* New York, Acad Pr, 1971.

Cold Spring Harbor Laboratory: *Symposia on Quantitative Biology,* Vol XXXIII: *Replication of DNA in Microorganisms.* Long Island, Cold Spring Harbor Laboratory, 1968.

Cold Spring Harbor Laboratory: *Symposia on Quantitative Biology,* Vol XXXVIII: *Chromosome Structure and Function.* Long Island, Cold Spring Harbor Laboratory, 1974.

Davidson, J. N.: *The Biochemistry of Nucleic Acids,* 7th ed. New York, Acad Pr, 1972.

Gross, J. D.: DNA replication in bacteria. *Curr Top Microbiol Immunol, 57:*39, 1972.

Moore, R. L.: Nucleic acid reassociation as a guide to genetic relatedness among bacteria. *Curr Top Microbiol Immunol, 64:*105, 1974.

Raacke, I. D.: *Molecular Biology of DNA and RNA: An Analysis.* St. Louis, Mosby, 1971.

Schekman, R., Weiner, A., and Kornberg, A.: Multienzyme systems of DNA replication. *Science, 186:*987, 1974.

Watson, J. D.: *The Double Helix.* New York, Atheneum, 1968.

Research Papers

Alloway, J. L.: Further observations on the use of pneumococcus extracts in effecting transformation of type *in vitro. J Exp Med, 57:*265, 1933.

Avery, O. T., MacLeod, C. M., and McCarty, M.: Studies on the chemical nature of the substance inducing transformation of pneumococcal types. Induction of transformation by a desoxyribonucleic acid fraction isolated from pneumococcus type III. *J Exp Med, 79:*137, 1944.

Chargaff, E.: Chemical specificity of nucleic acids and mechanism of their enzymatic degradation. *Experientia, 6:*201, 1950.

Fritsch, A., and Worcel, A.: Symmetric multifork chromosome replication in fast-growing *Escherchia coli. J Mol Biol, 59:*207, 1971.

Gates, F. L.: On nuclear derivatives and the lethal action of ultra-violet light. *Science, 68:*479, 1928.

Gilbert, W., and Dressler, D.: DNA replication: the rolling circle model. *Cold Spring Harbor Symp Quantit Biol, 33:*473, 1968.

Griffith, F.: Significance of pneumococcal types. *J Hyg* (Camb), *27:*113, 1928.

Goulian, M., Kornberg, A., and Sinsheimer, R. L.: Enzymatic synthesis of DNA, XXIV. Synthesis of infectious phage φX174 DNA. *Proc Natl Acad Sci USA, 58:*2321, 1967.

Gyurasits, E. B., and Wake, R. G.: Bidirectional chromosome replication in *Bacillus subtilis. J Mol Biol, 73:*55, 1973.

Hershey, A. D., and Chase, M.: Independent functions of viral protein and nucleic acid in growth of bacteriophage. *J Gen Physiol, 36:*39, 1952.

McKenna, W. G., and Masters, M.: Biochemical evidence for bidirectional replication of DNA in *E. coli. Nature, 240:*536, 1972.

Okazaki, R., Okazaki, T., Sakabe, K., Sugimoto, K., and Sugino, A.: Mechanism of DNA chain growth. I. Possibility of discontinuity and unusual secondary structure of newly synthesized chains. *Proc Natl Acad Sci USA, 59:*598, 1968.

Sugino, A., and Okazaki, R.: RNA-linked DNA fragments *in vitro. Proc Natl Acad Sci USA, 70:*88, 1973.

Watson, J. D., and Crick, F. H. C.: Molecular structure of nucleic acids. A structure for deoxyribonucleic acid. *Nature, 171:*737, 1953.

Watson, J. D., and Crick, F. H. C.: Genetical implications of the structure of deoxyribonucleic acid. *Nature, 171:*964, 1953.

CHAPTER 3

PROTEIN SYNTHESIS

Books and Review Articles

Cold Spring Harbor Laboratory: *Symposia on Quantitative Biology,* Vol XXVIII: *Synthesis and Structure of Macromolecules.* Long Island, New York, Cold Spring Harbor Laboratory, 1963.

Cold Spring Harbor Laboratory: *Symposia on Quantitative Biology,* Vol XXXI: *The Genetic Code.* Long Island, New York, Cold Spring Harbor Laboratory, 1966.

Goldberger, R. F.: Autogenous regulation of gene expression. *Science, 183:*810, 1974.

Haselkorn, R., and Rothman-Denes, L. B.: Protein synthesis. *Annu Rev Biochem, 42:*397, 1973.

Jacob, F., and Monod, J.: On the regulation of gene activity. *Cold Spring Harbor Symp Quant Biol, 26:*193, 1961.

Littsur, V. Z., and Inouye, H.: Regulation of tRNA. *Annu Rev Biochem, 42:*439, 1973.

Mandelstam, J., and McQuillen, K. (Eds.): *Biochemistry of Bacterial Growth.* New York, Wiley, 1968.

Nomura, M.: Bacterial ribosome. *Bacteriol Rev, 34:*228, 1970.

Reznikoff, W. S.: The operon revisited. *Annu Rev Genet, 6:*133, 1972.

Woese, C. R.: *The Genetic Code. The Molecular Basis for Genetic Expression.* New York, Har-Row, 1967.

Research Papers

Blasi, F., Bruni, C. B., Avitabile, A., Deeley, R. G., Goldberger, R. F., and Meyers, M. M.: Inhibition of transcription of the histidine operon *in vitro* by the first enzyme of the histidine pathway. *Proc Natl Acad Sci USA, 70:*2692, 1973.

Brenner, S., Jacob, F., and Meselson, M.: An unstable intermediate carrying information from genes to ribosomes for protein synthesis. *Nature, 190:*576, 1961.

Crick, F. H. C., Barnett, L., Brenner, S., and Watts-Tobin, R. J.: General nature of the genetic code for proteins. *Nature, 192:*1227, 1961.

Dickson, R. I., Abelson, J., Barnes, W. M., and Reznikoff, W. S.: Genetic regulation: the lac control region. *Science, 187:*27, 1975.

Gamow, G.: Possible relation between deoxyribonucleic acid and protein structures. *Nature, 173:*318, 1954.

Gilbert, W., and Müller-Hill, B.: Isolation of the *Lac* repressor. *Proc Nat Acad Sci,* USA, *56:*1891, 1966.

Holmquist, R., Jukes, T. H., and Pangburn, S.: Evolution of transfer RNA. *J Mol Biol, 78:*91, 1973.

Kim, S. H., Sussman, J. L., Suddath, F. L., Quigley, G. J., McPherson, A., Wang, A. H. J., Seeman, N. C., and Rich, A.: The general structure of transfer RNA molecules. *Proc Natl Acad Sci USA, 71:*4970, 1974.

Nirenberg, M., and Leder, P.: RNA codewords and protein synthesis. The effect of trinucleotides upon the binding of sRNA to ribosomes. *Science, 145:*1399, 1964.

Nirenberg, M. W., and Matthaei, J. H.: The dependence of cell-free protein synthesis in *Escherchia coli* upon naturally occurring or synthetic polyribonucleotides. *Proc Natl Acad Sci USA, 47:*1588, 1961.

Nishimura, S.: Minor components in transfer RNA: Their characterization, location and function. *Prog Nucleic Acid Res Mol Biol, 12:*49, 1972.

CHAPTER 4

MUTATIONS I

Books and Review Articles

Auerbach, C.: The chemical production of mutations. *Science, 158:*1141, 1967.

Doudney, C. O.: Ultraviolet light effects on the bacterial cell. *Curr Top Microbiol Immunol, 46:*116, 1968.

Drake, J. W.: *The Molecular Basis of Mutation.* San Francisco, Holden-Day, 1970.

Strauss, B. S.: DNA repair mechanisms and their relation to mutation and recombination. *Curr Top Microbiol Immunol, 44:*1, 1968.

Witkin, E. M.: Ultraviolet-induced mutation and DNA repair. *Annu Rev Genet, 3:*525, 1969.

Research Papers

Altenbern, R. A.: Marker frequency analysis mapping of the *Staphylococcus aureus* chromosome. *Can J Microbiol, 17:*903, 1971.

Demerec, M., and Latarjet, R.: Mutations in bacteria induced by radiations. *Cold Spring Harbor Symp Quant Biol, 11:*38, 1946.

Helling, R. B.: Selection of a mutant of *Escherichia coli* which has high mutation rates. *J Bacteriol, 96:*975, 1968.

Kelner, A.: Photoreactivation of ultraviolet-irradiated *Escherichia coli*, with special reference to the dose-reduction principle and to ultraviolet induced mutations. *J Bacteriol, 58:*511, 1949.

Newcombe, H. B.: Delayed phenotypic expression of spontaneous mutations in *Escherichia coli. Genetics, 33:*447, 1948.

Setlow, J. K., and Boling, M. E.: Ultraviolet action spectra for mutation in *Escherichia coli. Mutat Res, 9:*437, 1970.

Szybalski, W., and Bryson, V.: Genetic studies on microbial cross resistance to toxic agents. I. Cross resistance of *Escherichia coli* to fifteen antibiotics. *J Bacteriol, 64:*489, 1952.

CHAPTER 5

MUTATIONS II

Books and Review Articles

Balassa, G.: The genetic control of spore formation in bacilli. *Curr Top Microbiol Immunol, 56:*99, 1971.

Benveniste, R., and Davies, J.: Mechanisms of antibiotic resistance in bacteria. *Annu Rev Microbiol, 42:*471, 1973.

Elander, R. P.: Strain development in industrially important microorganisms. *Dev Indust Microbiol, 7:*61, 1966.

Elander, R. P.: Applications of microbial genetics to industrial fermentations. *In* Perlman, D. (Ed.): *Fermentation Advances.* New York, Acad Pr, 1969.

Hartman, P. E., and Roth, J. R.: Mechanisms of suppression. *Adv Genetics 17:*1, 1973.

Iino, T.: Genetics and chemistry of bacterial flagella. *Bacteriol Rev, 33:*454, 1969.

Zähner, H., and Maas, W. K.: *Biology of Antibiotics.* New York, Springer-Verlag, 1972.

Research Papers

Bryson, V., and Demerec, M.: Patterns of resistance to antimicrobial agents. *Ann NY Acad Sci, 53:*283, 1950.

Clark, A. J., and Margulies, A. D.: Isolation and characterization of recombination deficient mutants of *E. coli* K-12. *Proc Natl Acad Sci USA, 53:*451, 1965.

Cohen, A., Fisher, W. D., Curtiss, R., III, and Adler, H. I.: The properties of DNA transferred to minicells during conjugation. *Cold Spring Harbor Symp Quant Biol, 33:*635, 1968.

Davis, B. D.: Isolation of biochemically deficient mutants of bacteria by penicillin. *J Am Chem Soc, 70:*4267, 1948.

Iino, T.: Curly flagellar mutants in *Salmonella. J Gen Microbiol, 27:*167, 1962.

Leifson, E., and Palen, M. I.: Variations and spontaneous mutations in the genus *Listeria* in respect to flagellation and motility. *J Bacteriol, 70:*233, 1955.

Witkin, E. M.: Genetics of resistance to radiation in *Escherichia coli. Genetics, 32:*221, 1947.

CHAPTER 6

RECOMBINATION IN BACTERIA

Books and Review Articles

Achtman, M.: Genetics of the F sex factor in *Enterobacteriaceae. Curr Top Microbiol Immunol, 60:*79, 1973.

Adelberg, E. A., and Pittard, J.: Chromosome transfer in bacterial conjugation. *Bacteriol Rev, 29:*161, 1965.

Curtiss, R., III: Bacterial conjugation. *Annu Rev Microbiol, 23:*69, 1969.

Erickson, R. J.: New ideas and data on competence and DNA entry in transformation of *Bacillus subtilis. Curr Top Microbiol Immunol, 53:*149, 1970.

Jacob, F. and Wollman, E. L.: *Sexuality and the Genetics of Bacteria.* New York, Acad Pr, 1961.

Levinthal, M.: Bacterial genetics excluding *E. coli. Annu Rev Microbiol, 28:*219, 1974.

Notani, N. K., and Setlow, J. K.: Mechanism of bacterial transformation and transfection. *Prog Nucleic Acid Res Mol Biol, 14:*39, 1974.

Ravin, A. W.: The origin of bacterial species. Genetic recombination and factors limiting it between bacterial populations. *Bacteriol Rev 24:*201, 1960.

Sanderson, K. E.: Linkage map of *Salmonella typhimurium,* edition IV. *Bacteriol Rev, 36:*558, 1972.

Research Papers

Adelberg, E. A., and Burns, S. N.: Genetic variation in the sex factor of *Escherichia coli. J Bacteriol, 79:*321, 1960.

Arwert, F., and Venema, G.: Transformation in *Bacillus subtilis.* Fate of newly introduced transforming DNA. *Mol Gen Genet, 123:*185, 1973.

Cavalli, L. L., Lederberg, J., and Lederberg, E. M.: An infective factor controlling sex compatibility in *Bacterium coli. J Gen Microbiol, 8:*89, 1953.

Felkner, I. C., and Wyss, O.: Regulation of competence development in *Bacillus subtilis. Biochem Biophys Res Commun, 41:*901, 1970.

Goodgal, S. H.: Studies on transformations of *Hemophilus influenzae.* IV Linked and unlinked transformation. *J Gen Physiol, 45:*205, 1961.

Hayes, W.: Observations on a transmissible agent determining sexual differentiation in *Bacterium coli. J Gen Microbiol, 8:*72, 1953.

Lederberg, J., and Tatum, E. L.: Novel genotypes in mixed cultures of biochemical mutants of bacteria. *Cold Spring Harbor Symp Quant Biol, 11:*113, 1946.

Marmur, J.: A procedure for the isolation of deoxyribonucleic acid from micro-organisms. *J Molec Biol, 3:*208, 1961.

Oishi, M., and Cosloy, S. D.: Specialised transformation in *Escherichia coli* K-12. *Nature, 248:*112, 1974.

Riva, S., and Polsinelli, M.: Relationship between competence for transfection and for transformation. *J Virol, 2:*587, 1968.

Romig, W. R.: Infection of *Bacillus subtilis* with phenol-extracted bacteriophages. *Virology, 16:*452, 1962.

Saito, H., and Miura, K.: Preparation of transforming deoxyribonucleic acid by phenol treatment. *Biochim Biophys Acta, 72:*619, 1963.

Spizizen, J.: Transformation of biochemically deficient strains of *Bacillus subtilis* by deoxyribonucleate. *Proc Natl Acad Sci USA, 44:*1072, 1958.

CHAPTER 7

BACTERIAL VIRUSES AND PLASMIDS

Books and Review Articles

Bacterial Viruses

Adams, M.: *Bacteriophages.* New York, Interscience, 1959.

Arber, W.: DNA modification and restriction. *Prog Nucleic Acid Res Mol Biol* *14:*1, 1974.

Boyer, H. W.: DNA restriction and modification mechanisms in bacteria. *Annu Rev Microbiol, 25:*153, 1971.

Bradley, D. E.: Ultrastructure of bacteriophages and bacteriocins. *Bacteriol Rev, 31:*230, 1967.

Calendar, R.: The regulation of phage development. *Annu Rev Microbiol, 24:*241, 1970.

Echols, H.: Developmental pathways for the temperate phages: lysis vs. lysogeny. *Annu Rev Genet, 6:*157, 1972.

Herskowitz, I.: Control of gene expression in bacteriophage lambda. *Annu Rev Genet, 7:*289, 1973.

Padan, E. and Shilo, M.: Cyanophages—viruses attacking blue-green algae. *Bacteriol Rev, 37:*343, 1973.

Signer, E. R.: Lysogeny: The integration problem. *Annu Rev Microbiol, 22:*451, 1968.

Plasmids

Campbell, A. M.: *Episomes.* New York, Har-Row, 1969.

Clowes, R. C.: Molecular structure of bacterial plasmids. *Bacteriol Rev, 36:*361, 1972.

Meynell, E., Meynell, G. G., and Datta, N.: Phylogenetic relationships of drug-resistance factors and other transmissible bacterial plasmids. *Bacteriol Rev, 32:*55, 1968.

Meynell, G. G.: *Bacterial Plasmids. Conjugation, Colicinogeny, and Transmissible Drug-resistance.* London, MacMillan, 1972.

Mitsuhashi, S. (Ed.): *Transferable Drug Resistance Factor R.* Baltimore, Univ Park, 1971.

Novick, R. P.: Extrachromosomal inheritance in bacteria. *Bacteriol Rev, 33:*210, 1969.

Richmond, M. H.: Resistance factors and their ecological importance to bacteria and to man. *Prog Nucleic Acid Res Mol Biol, 13:*191, 1973.

Willetts, N.: The genetics of transmissible plasmids. *Annu Rev Genet, 6:*257, 1972.

Research Papers

Bacterial Viruses

Benzer, S.: On the topology of the genetic fine structure. *Proc Natl Acad Sci USA, 45:*1607, 1959.

Delbrück, M., and Bailey, W. T., Jr.: Induced mutations in bacterial viruses. *Cold Spring Harbor Symp Quant Biol, 11:*33, 1946.

Edgar, R. S., and Epstein, R. H.: The genetics of a bacterial virus. *Sci Am, 212(2):*70, 1965.

Hershey, A. D.: Mutation of bacteriophage with respect to type of plaque. *Genetics, 31:*620, 1946.

Hershey, A. D., and Rotman, R.: Genetic recombination between host-range and plaque-type mutants of bacteriophage in single bacterial cells. *Genetics, 34:*44, 1949.

Lanni, Y. T.: First-step-transfer deoxyribonucleic acid of bacteriophage T5. *Bacteriol Rev, 32:*227, 1968.

Lieb, M.: λ mutants which persist as plasmids. *J Virol, 6:*218, 1970.

Morse, M. L., Lederberg, E. M., and Lederberg, J.: Transduction in *Escherichia coli* K-12. *Genetics, 41:*142, 1956.

Notani, G. W.: Regulation of bacteriophage T4 gene expression. *J Mol Biol, 73:*231, 1973.

Pappenheimer, A. M., and Gill, D. M.: Diphtheria. *Science, 182:*353, 1973.

Sinsheimer, R. L., Knippers, R., and Komano, T.: Stages in the replication of bacteriophage φX174 DNA *in vivo. Cold Spring Harbor Symp Quant Biol, 33:*443, 1968.

Visconti, N., and Delbrück, M.: The mechanism of genetic recombination in phage. *Genetics, 38:*5, 1953.

Zinder, N. D., and Lederberg, J.: Genetic exchange in *Salmonella. J Bacteriol, 64:*679, 1952.

Plasmids

Hershfield, V., Boyer, H. W., Yanofsky, C., Lovett, M. A., and Helinski, D. R.: Plasmid ColE1 as a molecular vehicle for cloning and amplification of DNA. *Proc Natl Acad Sci USA, 71:*3455, 1974.

LeBlanc, D. J., and Falkow, S.: Effects of superinfection immunity on plasmid regulation following conjugation. In Krcmery, V., Rosival, L., and Watanabe, T. (Eds.): *Bacterial Plasmids and Antibiotic Resistance.* New York, Springer-Verlag, 1972.

Summers, A. O., and Silver, S.: Mercury resistance in a plasmid-bearing strain of *Escherichia coli. J Bacteriol, 112:*1228, 1972.

Tschäpe, H., and Rische, H.: The virulence plasmids of *Enterobacteriaceae. Z Allg Mikrobiol, 14:*337, 1974.

Watanabe, T., and Fukasawa, T.: "Resistance transfer factor," an episome in *Enterobacteriaceae. Biochem Biophys Res Commun, 3:*660, 1960.

CHAPTER 8

THE VIRUSES OF EUCARYOTES

Books and Review Articles

Baltimore, D.: Expression of animal virus genomes. *Bacteriol Rev, 35:*235, 1971.

Brown, R. M., Jr.; Algal viruses. *Adv Virus Res, 17:*243, 1972.

Fenner, F.: The genetics of animal viruses. *Annu Rev Microbiol, 24:*297, 1970.

Knight, A. C.: *Molecular Virology.* New York, McGraw, 1974.

Lemke, P. A., and Nash, C. H.: Fungal Viruses. *Bacteriol Rev, 38:*29, 1974.

Luria, S. E. and Darnell, J. E., Jr.: *General Virology,* 2nd ed. New York, Wiley, 1967.

McAuslan, B. P.: Virus-associated enzymes. *Life Sci, 14:*2085, 1974.

Shatkin, A. J.: Viruses with segmented ribonucleic acid genomes: Multiplication of influenza versus reovirus. *Bacteriol Rev, 35:*250, 1971.

Temin, H. M.: The RNA tumor viruses: Background and foreground. *Proc Natl Acad Sci USA, 69:*1016, 1972.

Research Papers

Benjamin, T. L.: Host range mutants of polyoma virus. *Proc Natl Acad Sci USA, 67:*394, 1970.

Mayor, H., Torikai, K., Melnick, J. C., and Mandel, M.: Plus and minus single stranded DNA separately encapsulated in adeno-associated satellite virions. *Science, 166:*1280, 1969.

Schonberg, M., Silverstein, S. C., Levin, D. H., and Acs, G.: Asynchronous synthesis of the complementary strands of the reovirus genome. *Proc Natl Acad Sci USA, 68:*505, 1971.

Stanley, W. M.: Isolation of a crystalline protein possessing the properties of the tobacco-mosaic virus. *Science, 81:*644, 1935.

Wittmann, H. G., and Wittmann-Liebold, B.: Protein chemical studies of two RNA viruses and their mutants. *Cold Spring Harbor Symp Quant Biol, 31:*163, 1966.

INDEX